How to Measure Anything

60
5
10
15
20
25
30
35
40
45
50
55

How to Measure Anything

The Science of Measurement

General Editor
Christopher Joseph

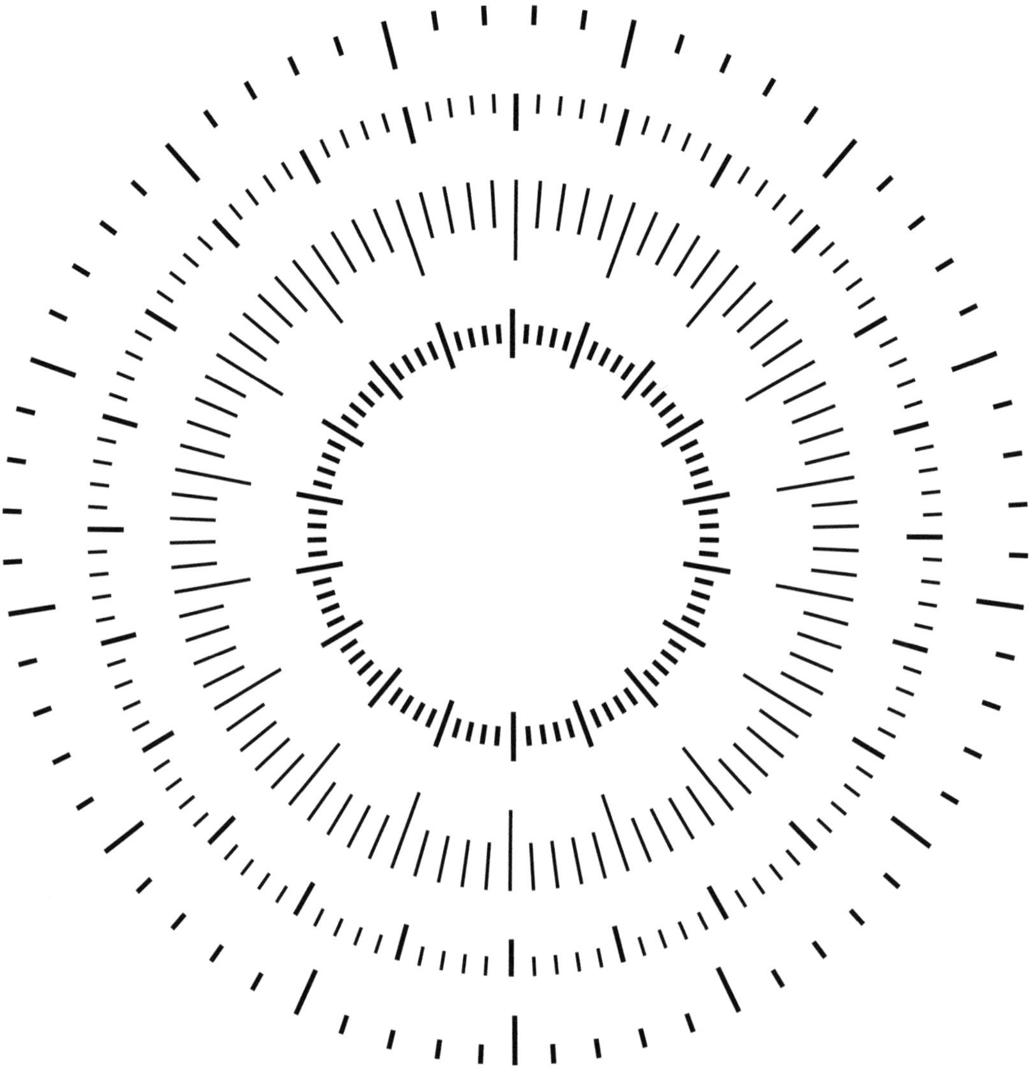

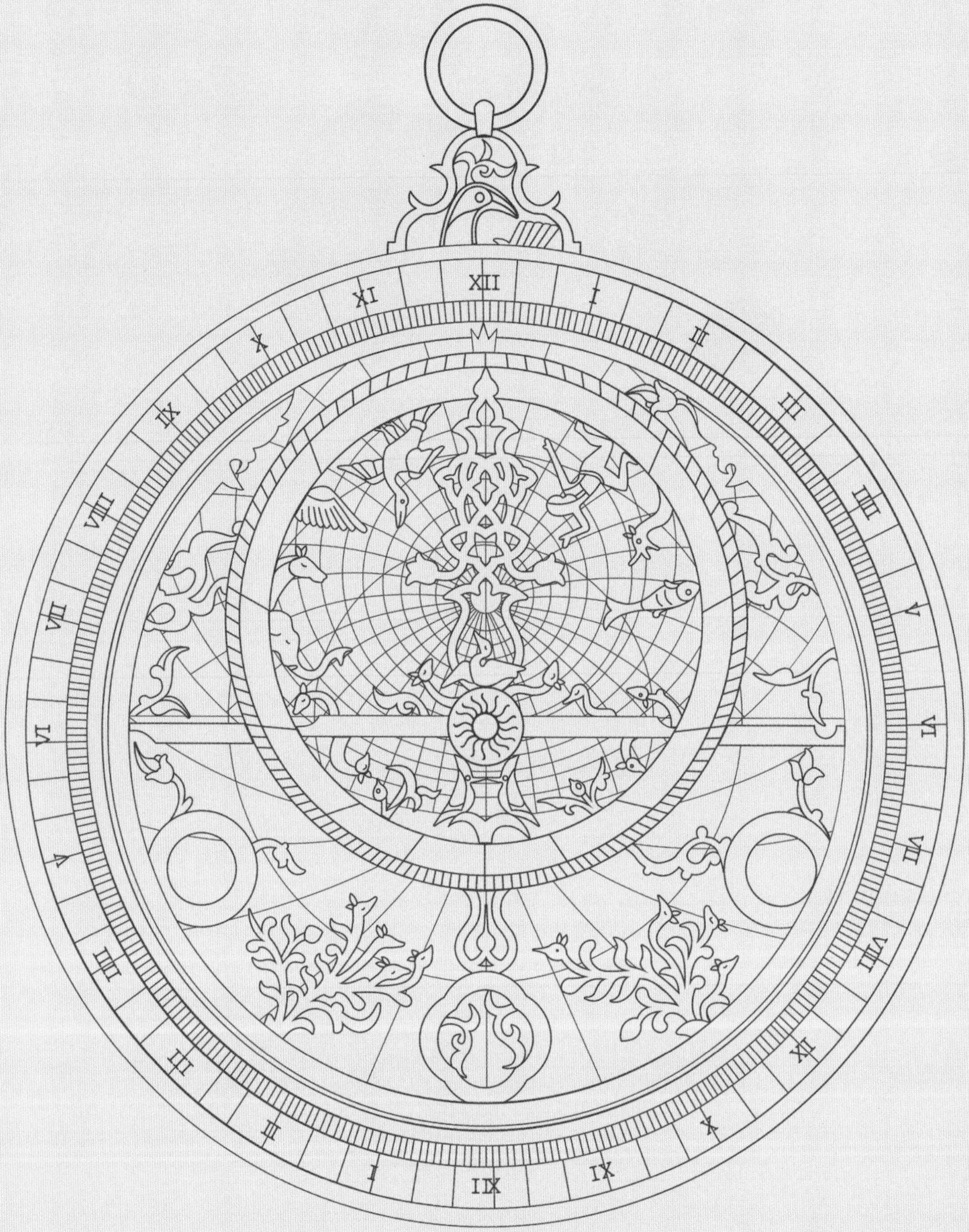

Introduction

Measurement, in one form or another, is one of mankind's oldest and most vital activities. Even before the dawn of civilization, relative measurement—"their tribe is bigger than ours"—was vital to the survival of any individual or group. The members of a hunter-gatherer society needed the concepts of "more," "less" and "enough" (enough time to get home before dark; enough food to ensure that no one will have to go hungry). With the creation of permanent settlements of ever-increasing size such estimation was no longer always sufficient. The increasing sophistication of language allowed comparisons to become ever more complex—enough for one person is not always enough for another.

The earliest historical record of a unit of measurement is the Egyptian cubit, decreed in around 3000 B.C.E. to be equal to the length of a forearm and hand plus the width of Pharaoh's palm. This is, of course, still rather flexible—my forearm is not the same length as yours, and since neither of us met Pharaoh, we cannot confidently say precisely how wide his palm was. It is not yet much of a step forward from measuring distance in hand-widths, paces or whatever other

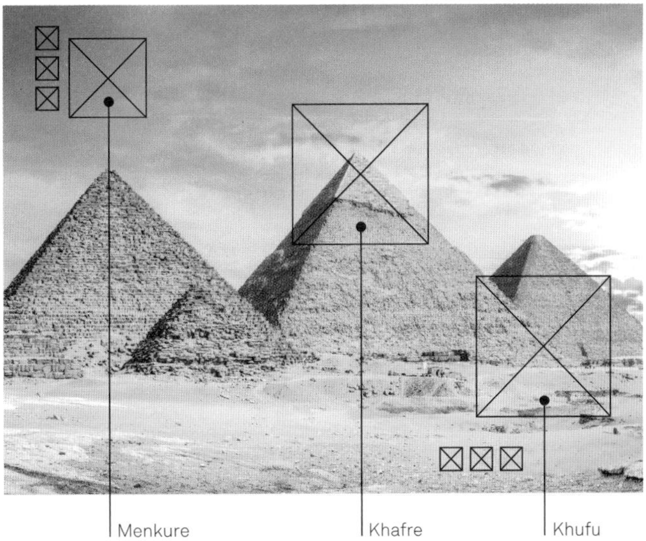

The pyramids at Giza in Egypt are geometric constructions of astonishing accuracy, even if the symbolic meaning of their proportions is not fully known. The largest of the three, Khufu's pyramid, remained the tallest manmade structure on Earth from its construction before 2500 B.C.E. until the 19th century —more than 4,300 years later.

approximation springs to mind.

By 2500 B.C.E., however, the complicated and rather imprecise definition had been simplified drastically: a cubit was the same length as the prototype "royal master cubit," a black marble rod some 52 cm in length. From this simple model it becomes possible to measure distances, areas, and volumes—and even masses in comparison with a volume of a specified substances such as gold or water.

While accurate measurements revolutionized aspects of human life ranging from craft and architecture to trade and transportation, the most important change they brought about was the possibility of science. Without accurate measurement—and equally accurate recording—no useful form of science, engineering or technology is really possible. Without accurate measurement, you couldn't be holding this book in your hands—a modern printing press is a huge piece of precision machinery.

One of the first application of this science was in measuring the heavens and refining calendars and timekeeping. Evidence of astronomical record-keeping dates back thousands of years—not just to literate civilisations like ancient Mesopotamia and Egypt, but also to the prehistoric stargazers of north-western Europe who erected monuments like Stonehenge with astonishing precision.

The influence of these early peoples and in particular those great engineers the Romans, can still be felt today, in an wide variety of ways. We owe to the Romans many of the names of our traditional units—and their language has also been plundered for scientific terms. "Ounce" and "inch" both derive from the Roman *uncia*, meaning a twelfth. Uses of *uncia* were not limited merely to weight and distance, however. Almost anything could sensibly be divided into 12 parts, and almost anything (at one time or another) was. To a Roman it made perfect sense to talk of *unciae* of

Although the famous bluestone slabs arrived at Stonehenge in the west of England from Wales in around 2500 B.C.E., the earliest similar construction on the site is believed to have been built some 600 years earlier.

In his astronomical treatise Almagest, one of the most influential books of Antiquity, Ptolemy compiled the astronomical knowledge of the ancient Greek and Babylonian world. This geocentric model of the solar system remained the generally accepted model in the western and Arab worlds until it was superseded by the heliocentric solar system of Copernicus.

Ptolemaic system

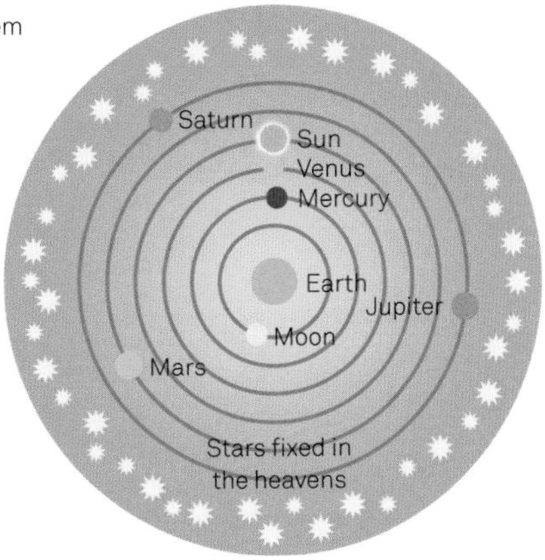

a cake, or an estate—perhaps even a business with ownership divided among several people.

For centuries after the fall of Rome, systems of customary units remained largely unchanged; there was, after all, little new to measure. Still, the feudal monarchs of Europe were interested in measurement for the same reason that any government is: if you do not know what someone owns or produces, how can you tax them on it?

As the Renaissance reinvigorated progress in art and science, however, interest in measurement regained some of its former vigor. Traditional units, however, continued in use—sometimes more different from one town to its neighbour than they were from one century to the next. Steadily, though, they became more accurate and reliable. Candles with hour markings were replaced by clockwork. While a clock in one town might tell a different time from that in a town 20 miles away, identical clocks could at least be relied upon to keep each hour to the same length.

Then came the French Revolution, and the great reform heralded by the introduction of the metric system. Though initially unpopular even in France, this system was the logical measuring accompaniment to the decimal Arabic numerals that had by then firmly supplanted Roman numbers throughout Europe. Decimal numbers made calculations of all kinds far more straightforward, as well as greatly simplifying much mathematical thinking. Trade and measurement, however, was another matter: 10 sheep were 10 sheep, but a pound was still 12 ounces. Conversions between different types of measurement were more complex still—if one pint of something weighed one pound, one gallon of it most definitely did not weigh one stone. The metric system supplanted all of this with a system of related units, distinctions of scale being made purely by the use of prefixes to indicate the magnitude of the number.

The standard unit forming the base of this system, from which all of the others were initially derived, was the meter. It was defined as one-ten-millionth of the distance from the Equator to the North Pole along a

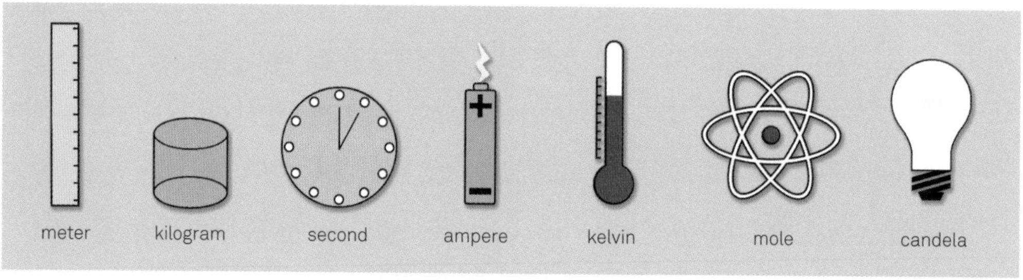

Although many other SI units have names, they can all be defined in terms of the seven base units.

meridian passing through Paris (in a steady curve at a theoretical sea level, rather than along the less-than-smooth surface of Earth). An enormous survey was carried out at great expense to calculate this distance as accurately as possible, and a platinum-iridium bar—the equivalent of Pharaoh's master cubit—cast as a permanent record. Carefully manufactured duplicates were distributed throughout Europe (and, ultimately, the world) to ensure that everyone was using the same meter.

As scientific equipment continued to improve, however, more accurate measurements revealed that not only did the platinum-iridium prototypes for the meter and other units such as the kilogram not match their intended definitions, but they might actually be changing as time passed. After some thought, the 19th-century committee decided to retain the prototypes—and therefore the current values of the units—rather than make a fresh attempt to match the definition.

By the 20th century, science and measurement had become inextricably entwined. While accurate measurement was a necessary prerequisite for good science, the continuing advancement of science provided an unending stream of new things to measure and new ways of measuring old things.

In the 1960s, the C.G.P.M. (Conférence Générale des Poids et Mesures), which meets every few years to discuss and resolve problems with the SI units and their use, decided enough was enough. It introduced sweeping redefinitions, with six base units, including the meter and kilogram, along with the kelvin, the second, the ampere and the candela. The useful chemical term mole became a seventh soon afterward. From these, everything else could be defined. Over time, the base units were carefully redefined in terms of reliably reproducible physical measurements (most recently in 2019) so that they could be tested and confirmed by anyone with the time, equipment and desire to do so.

Great care was taken to ensure that the units of the SI were what is called "coherent." This is best explained as the consistency of scale between units. Thus, for example, the pascal is coherent with the newton and the meter because a pressure of 1 pascal is equal to a force of 1 newton acting on an area of 1 square meter.

Another advantage of the system is that, if you want to describe a thousand million tonnes, it is not necessary to come up with a new word. You merely check the list of SI-approved prefixes and there you are: one gigatonne. Even so, by the late 20th century, as scientists discovered ever more about the universe, and wanted to measure ever larger—and ever smaller—things, the system needed expansion. Astronomers at one end of the scale and physicists, investigating subatomic particles, at the other were running out of prefixes. In the 1970s and 80s, therefore, the C.G.P.M. extended the list of approved prefixes at both ends of the scale.

Even those countries still using such traditional units and measures as the pint, the pound and the mile have long since redefined them in terms of their metric equivalents. Many other inventions of the 1790s have fallen by the wayside, but the early slogan of metrication, "for all people, for all time" seems increasingly appropriate.

Science and technology, of course, continue to investigate and invent new and different things, some of which can be measured in terms of standard SI units, while others require new inventions either out of conceptual necessity, or for simple convenience. We know of many different units used by one group of people or another over the millennia, and likely there were many more for which we have no record. Some have long and detailed histories, while for others we may have little more than a name, with no indication of what fixed value they possessed,

The values of many others varied wildly with place or time. Many ancient units are known only to historians; others are more widely known, if not always correctly understood. The talent, for example, is referred to frequently in the Bible, and everyone knows it is a sum of money. But the value of a coin, in the days of the Old Testament, was normally just the value of the precious metal that it contained—and a talent was a great sum of money indeed; at 25 kg, rather heavier than any practical coin.

Science is intimately bound up with the story of measurement; each drives the other forward. But measurement, either conscious or instinctive, is a part of every human activity—choosing the right color, drawing in perspective, correctly valuing a house for sale, or fitting the correct number of syllables in a line of poetry.

This book is intended to provide a friendly guide to the world of measurement, but it is necessarily only a whistle-stop tour. A complete list of all the units of which we have records (or even those to which we can provide a definite value) would be as large as any dictionary—even without including the definitions of (and differences between) the properties those units measure, and the instruments that record them. In compiling such a book as this, many things must be excluded as too rare to be worth defining or too common to need it. Our goal has been to produce something that is both a useful reference book—when you need one—and an entertaining read—when you don't. I hope that we have succeeded, and that nothing has been left out by accident.

How the entries work
Within each entry, units of measure are highlighted in bold type where the reader may wish to cross-refer to another entry (via the index) for further detail or explanation. The symbol 👁 indicates entries that are illustrated.

Earth and life sciences

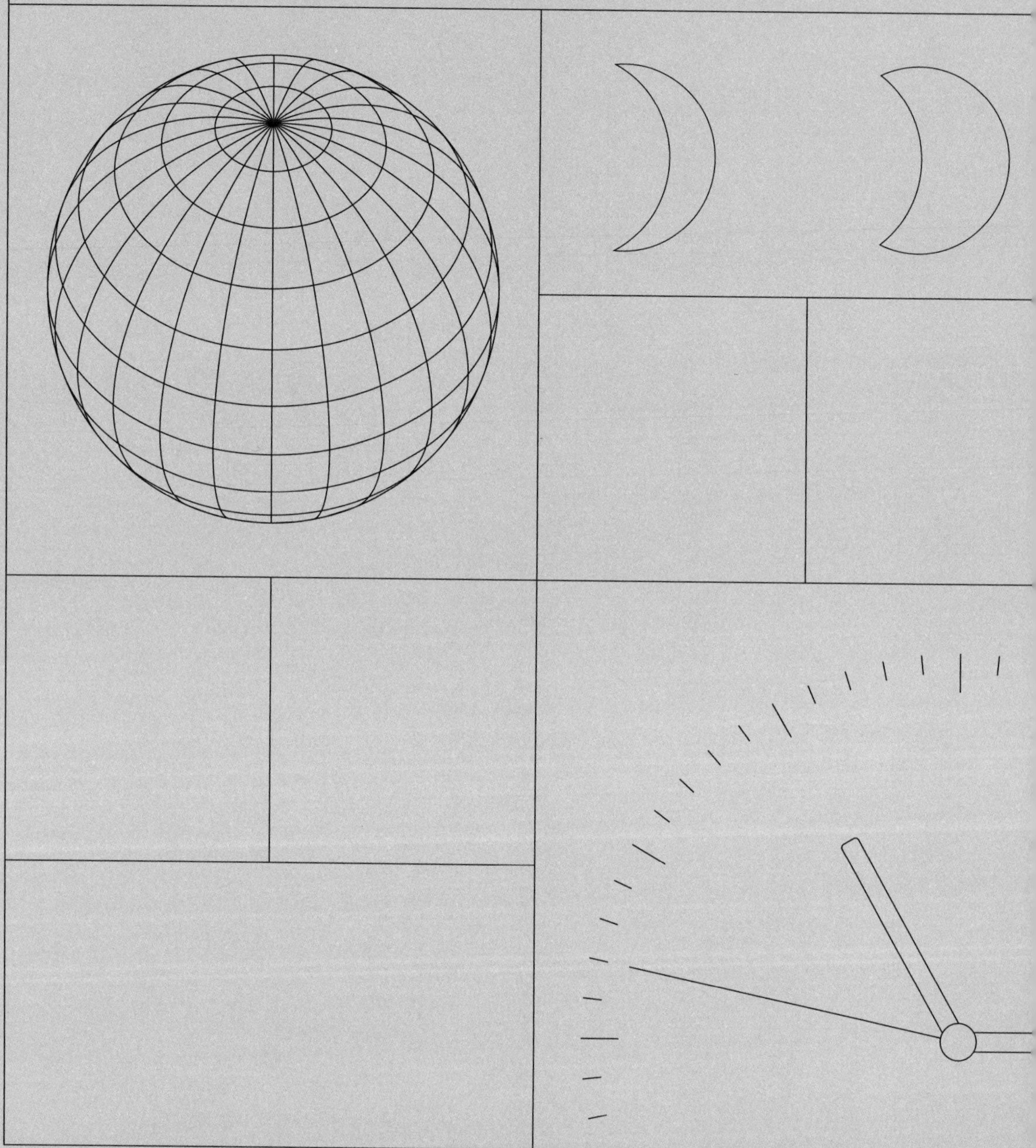

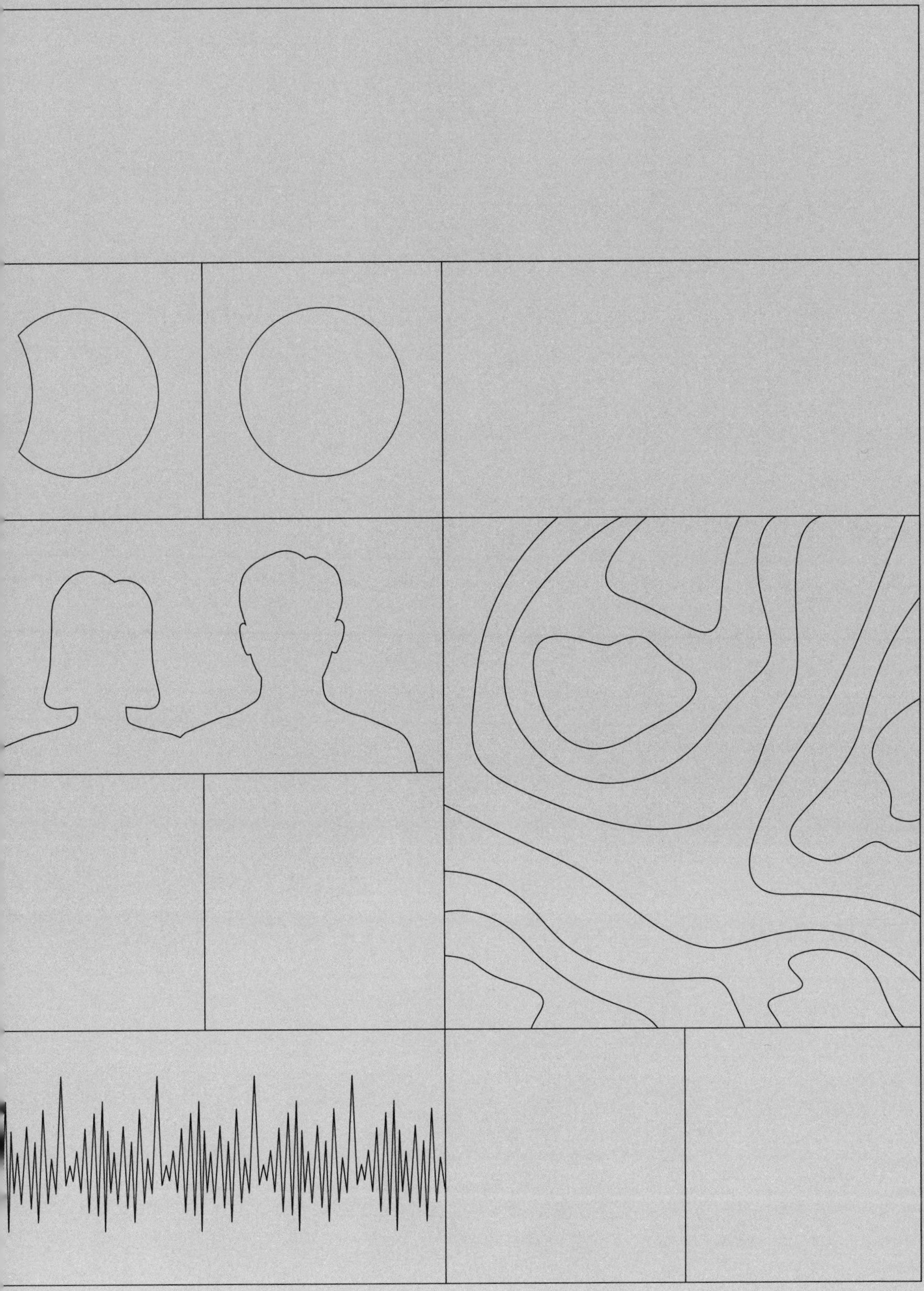

Astronomy and astrology

eccentricity
the extent to which a curved path (such as the orbit of a planet or comet around the sun) differs from a perfect circle, measured as the difference between its long and short axes, divided by the sum of those axes. An object with a highly eccentric orbit will come very close to the sun at one end of the orbit, with the other end being much farther away, whereas circular orbits have the sun at their center. For a closed orbit, eccentricity always has a value between zero (a perfect circle) and one.

apoapsis 👁
the point in its orbit at which the orbiting object is farthest from whatever it is orbiting. For an object orbiting the sun, apoapsis is called aphelion. An object in Earth orbit is at apogee when it reaches its greatest distance from the planet.

periapsis
the opposite of **apoapsis**; i.e., the point at which the distance between orbiting and orbited objects is smallest. For the sun's

As the eccentricity of a body's orbit increases, so does the difference between its apoapsis and periapsis.

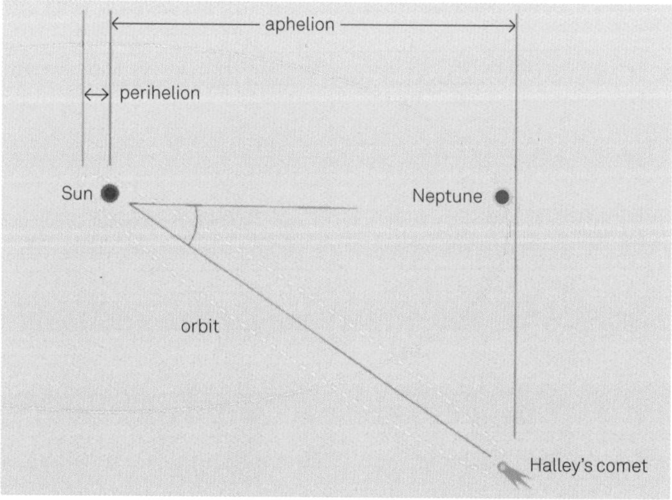

satellites, periapsis is called perihelion; for satellites in Earth orbit it is perigee. The moon's perigee is at around 359,000 km from Earth, some 42,000 km closer than its apogee. Halley's comet has its perihelion about 0.6 **AU** from the sun; the comet's aphelion is at 35.3 AU.

inclination
the angle between the orbital planes of two astronomical objects—usually, between an object and Earth. The orbital plane of an object is the flat surface that contains its entire orbit.

angular distance
the angle between two heavenly objects if they are projected as points onto a **celestial sphere** centered on Earth.

precession
the gradual change of direction of the axis of rotation of a rotating object (e.g., a planet, or spinning top) due to applied torque. For planets, precession is caused by gravitational tidal forces (those of the sun and moon, for Earth) pulling on the equator (since most planets are not quite spherical). Precession causes a slow change in the positions of the stars as measured in the equatorial **celestial coordinate** system. It also causes the equinoctial points to move westward along the ecliptic by around ⅚ of a second of arc each year, termed the precession of the equinoxes.

celestial sphere
an infinitely large imaginary sphere, centered on Earth. Every heavenly object (planets, stars, satellites, etc.) can be represented by a point on the celestial sphere, at the position where a line through the object from the center of Earth would touch the sphere.

ecliptic
an imaginary line around the celestial sphere marking the path followed by the Sun against the background stars in the course of a year. Due to Earth's axial tilt, the ecliptic is inclined at 23.43 degrees to the celestial equator (the projection of Earth's Equator onto the celestial sphere). Since the ecliptic is a projection of the plane of Earth's orbit onto the sky, it is generally also taken to be the plane of the entire solar system (since the other planets all orbit on relatively similar planes.

Different kinds of celestial coordinates are used by astronomers for different purposes.

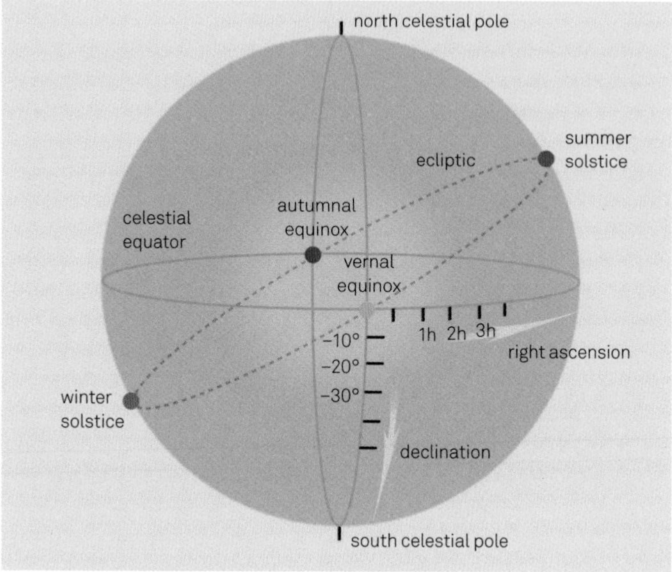

celestial coordinates

any system for mapping the position of objects on the celestial sphere. The main system used by professional astronomers is the "equatorial" system, which uses **declination** and **right ascension** to define the position of an object absolutely. Others, notably amateur astronomers and newspaper columnists, prefer the "horizontal coordinate" system, which uses **azimuth** and **elevation** to define the position of the object relative to the observer's location. The latter system is far easier to use instinctively, but the numbers it produces are entirely local—an object at zenith (i.e., 90° elevation) in Washington D.C., for example, will be much lower in Los Angeles. There are many other celestial coordinate systems, often designed for special purposes (e.g., one uses declination and ascension relative to the plane of the Milky Way galaxy for measurement of the galaxy itself).

declination

the **angular distance** of a celestial object from the projection of the Earth's Equator onto the celestial sphere. Positive declination means the object is north of the Equator; objects with negative declination lie south of it.

right ascension

the **angular distance** eastward from a vertical circle through the vernal equinox to the point where an object appears on the celestial sphere. Right ascension is normally measured in hours

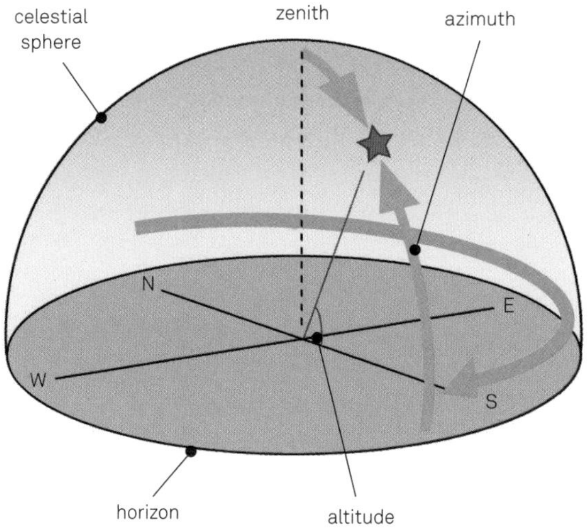

Azimuth and altitude are useful for finding positions relative to the observer.

(equal to 15° of arc), minutes (1⁄600 of an hour) and seconds (1⁄60 of a minute). The combination of declination and right ascension forms the "equatorial" set of **celestial coordinates**.

azimuth

the **angular distance** measured clockwise along the observer's horizon from the north point to the line from zenith to nadir passing through the object's apparent point on the celestial sphere.

elongation

the angular distance (in a direct line) between a celestial object and the sun, as seen from Earth. The inverse version (the angular distance between Earth and the sun, as seen from another celestial object) is called that object's phase angle.

orbital period

the length of time taken for an object to complete one orbit around whatever it may be orbiting; this is dependent on orbital velocity, eccentricity and its **apo**- and **periapses**.

orbital velocity (orbital speed)

the **speed** (not strictly **velocity**) at which a celestial object travels along its orbit.

Parallax causes objects at different distances to appear to move relative to one another, even when they are not actually doing so.

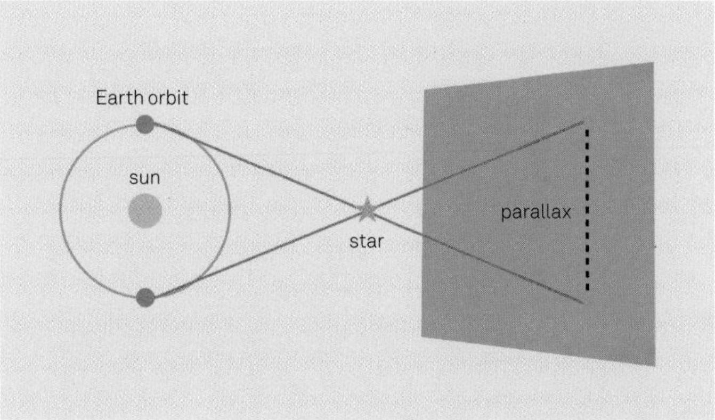

parallax

the apparent change in a celestial object's position depending on the location of the observer. In astronomy, this is usually caused by the movement of Earth, and is subdivided into diurnal parallax (caused by the Earth's daily rotation), annual parallax (caused by the Earth's movement as it orbits the sun) and secular parallax (caused by the movement of the solar system through space). The closer an object, the more it is affected by parallax—an effect observed in everyday life when roadside objects appear to move faster than distant ones viewed from a moving car.

proper motion

the angular movement of a star (or other object) on the celestial sphere. Proper motion is the portion of its actual movement that is perpendicular to the observer's line of sight.

radial velocity

the component of a celestial body's velocity directly along the observer's line of sight, usually calculated from the **shift** (red or blue) of the star's emitted light (or of the reflected light, for non-emitting bodies).

Hubble constant

the rate at which distant objects move apart from one another due to the expansion of space, measured as a function of their current distance. The Hubble constant is a constant only in that it applies to all objects at the present time; for much of the 20th century it was assumed that the expansion of space was entirely due to the big bang in which the Universe was created as a hot, compact state 13.8 billion years ago, and so would naturally slow down over time. However recent evidence suggests the rate of expansion is actually increasing due to a still-unexplained phenomenon called dark energy. The best measurement of the constant's current value is around 70 km per second per megaparsec.

escape velocity
the minimum speed needed to move from the surface of an object to a position beyond its gravitational field. Earth's escape velocity is just over 25,000 miles per hour (11,200 meters per second). It is not necessary for a vehicle to actually reach escape velocity in order to enter a stable orbit and experience "weightless" conditions.

Lagrangian point
in a gravitational system with two large bodies, the points in space where the gravitational fields of those two bodies cancel out. There are five points; two are stable (an object nearby will tend to drift towards them) and the other three are unstable (so that an object not perfectly centered there will tend to drift away).

geostationary orbit
an orbit over the equator of a planet at a height such that the orbital period is exactly equal to the period of rotation of the planet, meaning that a satellite in such an orbit remains stationary relative to the planetary surface underneath. For Earth, the altitude of geostationary orbit is 22,500 miles (36,000 km).

Roche limit 👁
the minimum distance from the parent planet's center at which a satellite can orbit the planet safely. Within the Roche limit, the satellite will, depending on its relative mass, either spiral rapidly to the ground or be torn apart, forming rings (such as those of Saturn). If a planet and its moon are of similar density, the Roche limit is around 2.5 times the planet's radius.

Schwarzschild radius
the minimum radius of an object with specified mass at which the surface **escape velocity** is lower than the speed of light. Any object that shrinks below the Schwarzschild radius for its mass becomes a black hole. Once such an object has collapsed, the Schwarzschild radius becomes its **event horizon**.

event horizon
a boundary from beyond which no information can reach an

The Roche limit varies with density; the denser the moon is relative to the planet, the closer it can come without being destroyed.

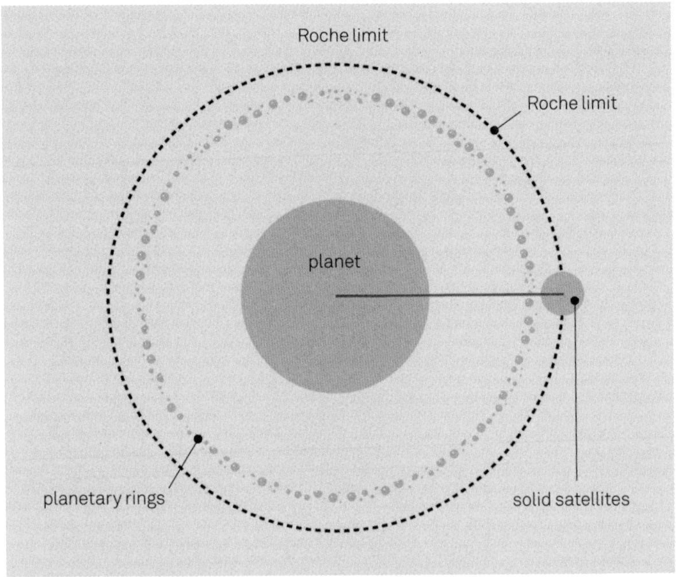

external observer, most commonly experienced around black holes where the escape velocity at the boundary exceeds the speed of light and other electromagnetic radiation.

planisphere
a circular star chart showing all the celestial objects that can possibly be seen from a particular **latitude**, combined with an overlay marked with dates and times. The overlay is positioned on the chart according to the current date and time, with adjustment for the precise latitude and **longitude**, and allows the calculation of horizontal **celestial coordinates** (i.e., **azimuth** and **elevation**) for any visible object.

bifilar micrometer
a simple measuring instrument for finding the angular separation of two objects viewed through the same telescope using a pair of extremely fine parallel wires. Sometimes called a filar micrometer.

heliometer
an instrument for measuring the **angular distance** between two objects, using a movable lens to generate a double image. The first image of one star is then aligned with the second image of the other by adjusting the lens, and the angular distance between them calculated from the adjustment required. It was originally designed for measuring the diameter of the sun, hence the name.

jansky (Jy)
a unit of received power from a source of electromagnetic radiation, primarily used for the strength of signals received by radio telescopes. One jansky is equal to 10^{-26} watts per square meter per hertz of channel width. Thus, a signal with a channel width of 1 MHz would need 1,000 times as much absolute power as one with a channel width of 1 kHz to have the same rating in janskys.

shift (redshift, blueshift)
the change in the apparent wavelength of light (and other radiation) emitted by distant stars as seen from Earth. The two main causes of shift are relative movement (the **Doppler effect**) and the expansion of intervening space (though the latter is only significant over great distances). Redshift happens when an object is moving away from us; blueshift when it is coming closer. Extreme gravitational fields can also generate shifts in the wavelength of light passing through them.

spectral class
a measure of the temperature and chemical composition of a star, based on the intensity of light it emits at different **wavelengths**. As with solar flares (*see* p. 24), the class consists of a letter followed by a number between 1 and 9, sometimes with further information (such as a **luminosity class**) after the number. Most stars fall into one of the letter-classes (in order from hottest to coldest) O, B, A, F, G, K and M, but some unusual stars are assigned other letters. The numbers are a linear scale from hottest to coldest within each class, so an A4 star is hotter than an A6 one. The sun's spectral class is G2.

luminosity class
a measure of a star's brightness, usually appended to its spectral class. Indicated with Roman numerals, on a scale theoretically from 1 to 7, though classes 6 and 7 are now rarely used. I indicates super-giants (the largest, brightest stars), while V is the much more numerous main sequence dwarf stars—such as the sun.

magnitude
a measure of a star's brightness. The apparent magnitude of a star (or other celestial object) is its brightness as seen from Earth, defined on a geometric scale with lower numbers indicating brighter stars. The dimmest objects visible from Earth with the naked eye have a magnitude of six, and a reduction of one

A simple model of how a heliometer works.

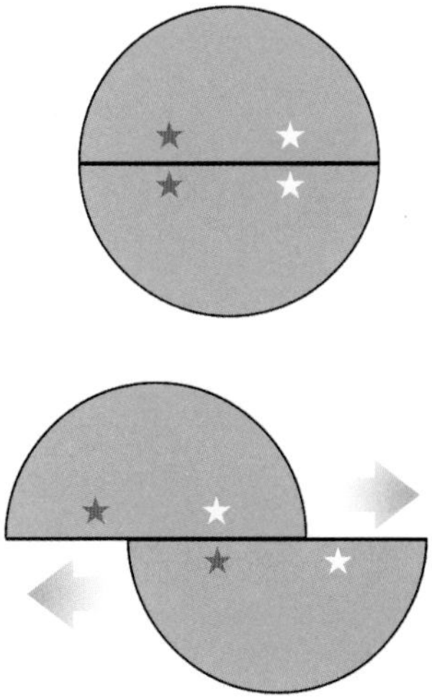

whole magnitude on the scale makes an object 2.51 times brighter. The absolute magnitude of a star is the apparent magnitude it would have if seen from a distance of exactly 10 parsecs.

sunspot number (Wolf sunspot number; R)

a measure of the sun's surface activity as indicated by the presence of sunspots. Equal to the sum of the total number of visible sunspots plus 10 times the number of sunspot groups, multiplied by a factor (usually between 0 and 1) that depends on the position and type of the telescope used to make the observation.

solar flare intensity scale

a measure of the X-ray energy of a solar flare, and therefore the strength of the resulting disruption to radio and satellite communications. (The strongest flares can affect electronic equipment on the ground, including power transmission networks.) Solar activity is classified with a letter (A, B, C, D, M or X) and a number (normally from 1 to 9); A, B and C are the normal surface behavior of the sun and class D flares generally have no effect on Earth. Each letter class is 10 times more powerful than the previous one, with the numbers representing a linear scale within each class. Thus, an X3 flare is six times the power of an M5 flare, which is in turn five times the power of an M1 flare. The most powerful solar flare recorded to date occurred on November 4, 2003, with a rating of X28.

The Torino scale rates the danger posed by an object based on both the likelihood of collision and the amount of damage it would cause.

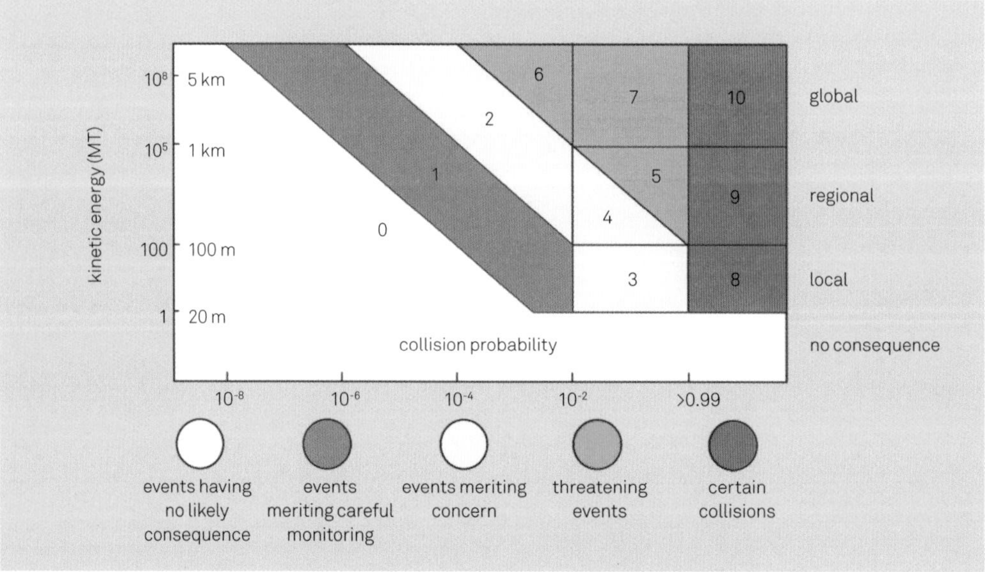

Torino impact scale 👁

a measure of the threat to human life posed by a comet, asteroid or other body approaching Earth. Events are rated on a scale of 0 to 10, subdivided into five classes. "Events having no likely consequence" (rating 0), and "events meriting careful monitoring" (rating 1) are low threat categories. "Events meriting concern" (ratings 2 to 4) are those where a collision is considered unlikely, and would only be likely to cause relatively local damage. "Threatening events" (ratings 5 to 7) are close encounters with significant possibilities of causing widespread (or even global) destruction, and ratings 8 to 10 are "certain collisions," ranging from relatively minor impacts causing purely localized devastation to global catastrophes capable of wiping out life on Earth.

radiant

the point on the **celestial sphere** from which a meteor shower or similar phenomenon appears to originate when observed from the Earth's surface. (It is actually a perspective effect of parallel lines, identical to the apparent narrowing of a straight road as it heads into the distance.)

epoch

in astronomical terms, a precise instant of time used as a point of reference for recorded data (e.g., the position of a star or planet in the sky). The combination of known starting time and location, and known movements, allows the calculation of the object's location at any other point in time. Where there is then a difference between the calculated and measured data, it may be used to correct assumptions or locate previously unknown objects whose gravity has affected the movement of the known object.

astronomical unit (AU) 👁

a unit of distance, small by astronomical standards, equal to 149.6 Gm (93 million miles or 150 million km). Roughly equal to the mean distance from Earth to the sun, the AU is actually defined as the radius of a perfectly circular orbit with the same period as Earth.

light-year (ly)

the distance that light will travel through a vacuum in one year, equivalent to 5.879 trillion miles, 9.461 trillion kilometres, or 63,241.1 AU . By extension, light minute, light second and so on are the distances light travels in the specified time.

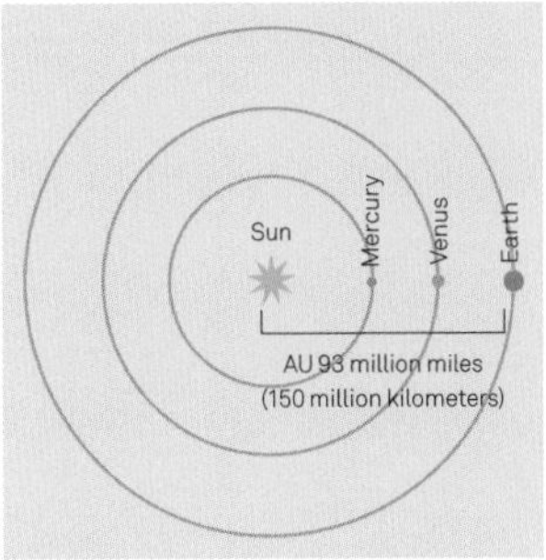

The Earth's orbit has an average radius of 1 AU. The average radius of Venus's orbit is about 0.7 AU, and for Mercury, about 0.38 AU.

parsec (pc)
the standard astronomical unit of distance, defined as the radius of a circle in which 1 arc-second of circumference is 1 **AU** in length. This means that a star at exactly 1 parsec distance would show a parallax shift of 1 arc second (1⁄3600th of a degree) as measured from Earth. One parsec is equal to 3.26 light-years, or just over 19.26 trillion miles. Proxima Centuri, the nearest star to our solar system, is 1.29 parsecs (4.22 light-years) away.

Hubble length (LH)
the maximum distance that an object can be from Earth and still be visible. It is the number of light-years equal to the age of the universe in years, since light from more distant objects cannot have reached us in that time. Due to the expansion of the Universe, objects 1 LH away from Earth are moving away from us at the speed of light.

solar mass
a unit of **mass** (not weight), used to express the mass of stars being studied by astronomers. Defined as 1.989×10^{30} kg (2.2×10^{27} "short" **tons**), the approximate mass of Earth's sun.

Jupiter mass
literally the **mass** of the planet Jupiter, used as a unit of mass by astronomers when discussing the planets in orbit around other stars. Approximately 1.9×10^{24} metric tonnes, or just under a thousandth of the mass of the sun.

Earth mass
the **mass** of our homeworld, just under 6×10^{21} metric tonnes (6.6×10^{21} federal or Customary tons—one Jupiter is around 315 Earths). Used as a comparative unit by astronomers looking for similar planets in other star systems.

moon, phases of
the change in shape of the visible part of the moon as its phase angle changes. The visible proportion of the moon's surface increases as the phase angle decreases, with full moon when the phase angle is very close to zero and a lunar eclipse when it is exactly zero. Planets, viewed through a telescope, have phases in exactly the same way.

The moon is referred to as waxing as it grows from new to full, and then waning as the visible area decreases again. A gibbous moon is one that is more than half-full—whether waxing or waning.

synodic period

the time taken for an object to return to the same position in the sky as seen from Earth (relative to the sun). The synodic month, for example, is the exact time between two consecutive full moons.

solar cycle

the length of time between two consecutive periods of maximum solar activity (when sunspots and **solar flares** are most common). The average period is approximately 11 years but cycles as short as nine and as long as 14 years have been known.

saros

a unit of time used for the prediction of eclipses, equal to exactly 223 synodic months, or—using the **Gregorian calendar**—18 years, 11 or 10 days (depending on the number of leap years) and 7.4 hours. After this length of time, Earth, sun and moon have returned to the same positions relative to one another. The extra hours, however, mean that each eclipse will be visible from a different part of the Earth's surface, returning to approximately its original position every three saros.

platonic year (great year)
the time taken for the Earth to make one complete precessional rotation (i.e., for the **precession** of the equinoxes to return the equinoctial points to their original positions on the ecliptic). It equals approximately 25,800 Earth years.

galactic year (cosmic year)
the time taken for the sun to make one complete orbit of the Milky Way galaxy. Generally reckoned to be approximately 225 million Earth years.

age of solar system
The time since significant solid clumps of material began to form from the pre-solar nebula—a vast cloud of gas and dust that provided the solar system's raw materials. Radio-isotope dating of meteorites and preserved minerals from Earth's most ancient rocks date its formation to around 4,570 million year ago.

age of the universe
the time calculated to have elapsed since the "Big Bang." The current best guess—based on estimates of the Universe's current size and rate of expansion among other sources—is 13.8 billion years.

zodiac
a set of 12 constellations, all approximately on the ecliptic, used by astrologers to indicate the relative positions of the various planets (along with the moon and sun). Modern astrology tends to work by dividing the ecliptic into 12 equal parts, while astrologers in ancient times used the actual boundaries of the constellations (which are not all of equal width). The 12 signs of the western zodiac are Aries, Taurus, Gemini, Cancer, Leo, Virgo, Libra, Scorpio, Sagittarius, Capricorn, Aquarius and Pisces, but other astrological systems have different zodiacal signs with often entirely contradictory meanings attached to them. Due to **precession**, the constellations have moved relative to the ecliptic since the zodiac constellations were first defined, and the sun spends a small amount of each year passing through a thirteenth constellation, Ophiuchus.

age (of Pisces, Aquarius, etc.)
an astrological period of time. Depending on the definitions used, this is either 2,100 years (the time the **precession** of the equinoxes takes to move 30° round the ecliptic), or the period for which the vernal equinox actually occurs within a particular zodiacal sign.

24.09 - 22.11
23.11 - 22.12
23.12 - 20.01
21.01 - 19.02
20.02 - 20.03
21.03 - 20.04
21.04 - 21.05
22.05 - 21.06
22.06 - 22.07
23.07 - 23.08
24.08 - 23.09
24.09 - 23.09

ascendant
the astrological sign of the zodiac that was rising over the horizon at the time and place of a person's birth.

midheaven
the astrological sign of the zodiac that was closest to **zenith** at the time and place of a person's birth. Since the signs of the zodiac are not of precisely equal widths, it is usually but not always the sign three ahead of the ascendant for the same time and place.

house
a set of 12 divisions on a person's horoscope. Depending on the astrologer, these may be of even or varying widths, assigned to a zodiacal sign or relative to the precise moment of birth.

opposition
a term used to describe the position of two objects that have ecliptic longitudes 180° apart.

conjunction
the opposite of **opposition**. Two planets (or a planet and the sun) are said by astrologers to be in conjunction when they appear close to one another in the sky.

Chinese astrological cycle
a more complex arrangement than its western counterpart, Chinese astrology has 12 animal signs (though these are not strictly a zodiac) and five "elements" (similar to the four elements of western alchemy). These combine to give a cycle of 60 years.

Distance

Some common metric prefixes. These are placed before units of metric measurement to indicate that the unit has increased by a certain factor.

Prefix	*Abbrev*	*Factor*
kilo-	k	10^3
hecto	h	10^2
deka-/deca-	da	10
—	—	1
deci-	d	10^{-1}
centi-	c	10^{-2}
milli-	m	10^{-3}
micro-	??	10^{-6}
nano-	n	10^{-9}

metric mile
a unit of length mostly used in athletics. A metric mile is different from an imperial, or statute, mile. Principally used as a race distance, it is generally accepted to be equivalent to 1,500 meters, or approximately 0.932057 imperial miles. Occasionally it is applied to a distance of 1,600 meters (much closer to an imperial mile); this usage is common in U.S. high-school races.

kilometer (km)
a unit of length equal to 1,000 meters, or 0.621371 imperial miles. The kilometer is the largest multiple of the meter in common usage.

meter (m)
the basic unit of length in the metric system. A meter is made up of 100 centimeters, or approximately 39.37 inches. The word derives from the Greek μετρον (*metron*). According to the Conférence Générale des Poids et Mesures, it is currently accurately defined as the distance light travels in a vacuum in $\frac{1}{299,792,458}$ seconds. This may change in the future as scientists make more accurate estimates as to the speed of light.

centimeter (cm)
a unit of length equal to $\frac{1}{100}$ of a meter or approximately 0.39 inches.

millimeter (mm)
a unit of length equal to $\frac{1}{1000}$ of a meter, or approximately 0.039 inches.

micrometer (µm, micron)
a unit of length equal to one-millionth of a meter, or approximately 0.00004 inches.A micrometer is also the term for an instrument used to measure small distances or thicknesses.

nanometer (nm)

a unit of length equal to one-thousand-millionth of a meter, or approximately 0.00000004 inches. The nanometer is used, among other things, to measure wavelengths, such as those of visible light, gamma rays and ultraviolet radiation.

angstrom (Å)

a unit of length equivalent to 10^{-10} meters, or 0.1 nanometers. The unit was first used by the Swedish physicist Anders Jonas Ångström (1814–74) to enable him to describe the solar spectrum. Although it is still occasionally used to describe the radii of atoms, which are between 0.25 and 3Å, it is more usual now to define such lengths in terms of nanometers.

femtometer (fm)

a unit of length equivalent to 10^{-15} meters, sometimes referred to by physicists as a fermi. The femtometer is used to measure the size of the nuclei of atoms, as well as that of sub-nucleic particles—protons and neutrons are approximately 2.5 femtometers in diameter.

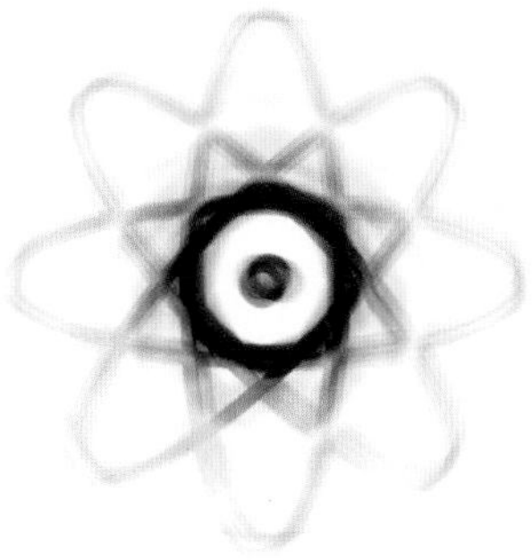

The diameter of nuclei of atoms is generally measured in femtometers.

Planck length (lp)

a unit of natural length used in quantum physics. First defined by German theoretical physicist Max Planck (1858–1947), it is the smallest unit of length defined by modern theoretical physics, and equivalent to 16.16 trillion trillion trillionths of a meter.

inch (in)

an imperial unit of length equivalent to 2.54 centimeters, or 1⁄36 yard. The inch is a very ancient measurement. It is thought that it was first defined as the distance between the tip of the thumb and the knuckle of the thumb. Indeed, in certain languages, the words for inch and thumb are very similar. The U.S. survey inch is fractionally larger than the standard imperial inch, the difference between the two only becoming significant when they are used to describe distances of many thousands of kilometers.

foot (ft)

an imperial unit of length equivalent to 12 inches, or 30.48 centimeters. It is popularly thought that the foot was originally defined by the length of a human foot. Human feet are generally smaller than 12 inches in length, however. In the U.S. it is more properly referred to as the international foot, to distinguish it from the U.S. survey foot (*see* **inch**).

yard (yd)
an imperial unit of length equivalent to 3 feet or 0.9144 meters. The yard has been standardized in a number of ways throughout history. Legend has it that Henry I of England defined it as the distance between the tip of his thumb and the tip of his nose with his arm outstretched.

rod
an imperial unit of length, traditionally employed to measure land. Equivalent to 16.5 feet or 5.03 meters, the rod is now only really used as a unit in North America, and it is also sometimes referred to as a perch (as distinct from a **perche**) or a pole. Historically the rod was a much less specific measurement.

perche
a unit of length and area, or both, having different definitions in different countries. The *perche* is quite distinct from the U.S. perch (see **rod**), and is defined in specific countries as follows: Canada, 231.822 inches, the Seychelles, approximately 6.497 meters; Switzerland, 3 meters; Belgium, 6.5 meters. In pre-metric France the *perche* was an important land measurement which had a number of different definitions according to the region.

chain
a non-metric unit of length used by surveyors. More properly known as Gunter's chain, and most commonly used in U.S. public land surveys, it is equivalent to 22 yards or 20.1168 meters, and is divided into 100 links. In Scotland and Ireland, Gunter's chain is of different, much shorter, lengths: 8.928 inches in Scotland, 10.08 inches in Ireland. There are two other types of chain, Ramden's and Rathborn's, but these are much less common. In Cyprus, a chain is 8 inches long.

furlong
an imperial and U.S. customary unit of length equivalent to 660 feet or 201.168 meters (one eighth of a mile). The word furlong derives from the Old English words *furh* (furrow) and *lang* (long), and it historically referred to the length of a furrow in a common field of 10 acres. The furlong's use is now largely restricted to British horse racing.

mile
an imperial unit of length also known as the international or statute mile. It is equivalent to 1,760 yards or approximately

1,609 meters. The nautical mile (also known as the "Admiralty mile"), however, is equivalent to 1,853 meters, and is used to navigate at sea and in the air. The statute mile was defined by the English Queen Elizabeth I as 8 furlongs in 1593, but the word itself derives from the Latin *mille passus* or **Roman mile**).

league
a historical unit of distance by land. Originally defined as the distance a person or a horse can walk in an hour, it became an accepted length of roughly 3 miles around the 16th century, though with several regional variations.

cable
a nautical unit of distance with a number of varying definitions. It is most commonly accepted as $\frac{1}{10}$ of a nautical mile, or 185.3 meters. Historically, however, a cable is 100 fathoms, or 182.88 meters. In the U.S. Navy, a cable is 120 fathoms, or 219.456 meters, whereas in the British Royal Navy, a cable is 608 feet, or 185.3184 meters.

fathom
a historical nautical unit of distance. A fathom is equivalent to 6 feet or 1.8288 meters, although it was originally deemed to be the width across a man's outstretched arms.

span
a unit of length equivalent to 9 inches or 22.86 centimeters. Historically it was taken to be the distance between the tips of the thumb and little finger when outstretched.

hand (hh)
a unit of length equivalent to 4 inches or 10.16 centimeters. Originally considered to be the breadth of a hand, it is now only used to measure the height of horses. The abbreviation hh stands for "hands high."

cubit
a very ancient unit of length used by several civilizations. The word derives from the Latin *cubitum*, meaning elbow, and was approximately equal to the length of a man's forearm. The Roman cubit was a length of about 44.35 centimeters, but the cubit also existed as a unit of measurement in the ancient kingdoms of Babylonia and Egypt. The Babylonian—or Sumerian—cubit was 51.72 cm, and this is the earliest known standard measurement of

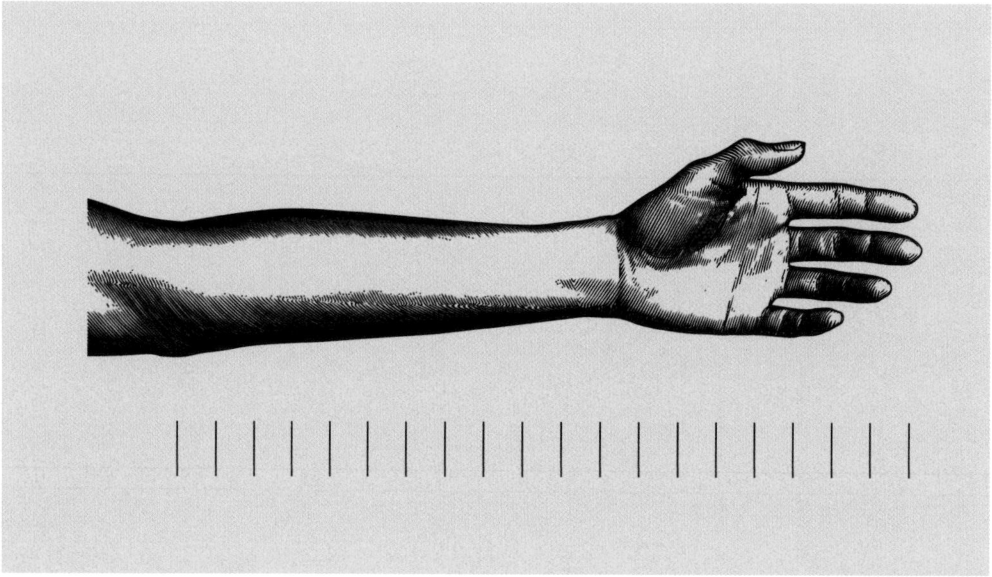

The cubit is one of the most ancient of all units of distance, but the word has been used to represent different distances throughout history and around the world.

length. The Egyptian cubit is thought to have been nearer 52.4 centimeters. The cubit was also used as a unit of measurement equivalent to 18 inches in the British Commonwealth, but it has now fallen out of use.

stadium

an ancient Roman or Greek unit of length. The English word stadium derives via Latin from the Greek σταδιον (*stadion*), and this unit of measurement corresponded to about 606 feet (185 m). The oval arenas in which the Greeks and Romans held their sports were generally built to this length, and so the unit of measurement gradually became the word for the arenas themselves.

Roman mile

an ancient Roman unit of length from which the modern statute mile is derived. The Latin for the Roman mile was *mille passus*, meaning "a thousand paces." It is accepted that the Roman mile corresponded to approximately 1,485 meters.

marathon

a long-distance running race. The marathon commemorates the myth that after the Greeks overcame the Persians in battle in 490 B.C.E., a messenger ran from Marathon to Athens—a distance of about 22 miles—with the news. (In fact, the Greek historian Herodotus gives an even earlier version of the story, in which the messenger Pheidippides ran not 22 but 150 miles from Athens to

Sparta to seek help before the battle.) The earliest marathon race was held at the first modern Olympic Games in 1896, but the length was not standardized by the International Olympic Committee until 1924, when it was set at 26 miles 385 yards, or 42.195 kilometers.

standard gauge
a standardized distance between the two rails of a railway track. Approximately 60 percent of the world's railway tracks are built to this gauge, which is 4 feet 8½ inches or 1.435 meters. This gauge was standardized in Britain by the Gauge Act of 1846—previously railway tracks had been set at the slightly narrower width of 4 feet 8 inches. Broad gauge, which is 7 feet ¼ inch or 2,140 mm wide, is used by high-speed trains in continental Europe.

caliber
the internal diameter of a gun barrel, or the diameter of a bullet or other piece of ammunition. Gun calibers can be described in inches (e.g. 0.44) or millimeters (e.g. 9 mm).

bore
in most uses, the internal diameter of a cylinder, measured in inches or millimeters. However when referring to shotguns (and some other weapons), the bore is the number of solid rounds for the weapon that can be made from 1 lb of lead. Hence a 4-bore shotgun is a more formidable weapon than a 12-bore.

calipers
an instrument for measuring dimensions. Resembling a pair of compasses, a set of calipers consists of two hinged legs with sharp points. These points are either directed outwards (to measure internal distances) or inwards (to measure external distances).

vernier scale
a moveable scale on the main scale of a measuring device. Devised by the French mathematician Pierre Vernier (1580–1637), a vernier scale enables the user to obtain fractional parts of the subdivisions on the main scale of a measuring device. Vernier scales can be found on, for example, barometers, sextants, calipers and micrometers.

odometer
an instrument for measuring the distance traveled by a wheeled object. Although most commonly encountered in cars, the

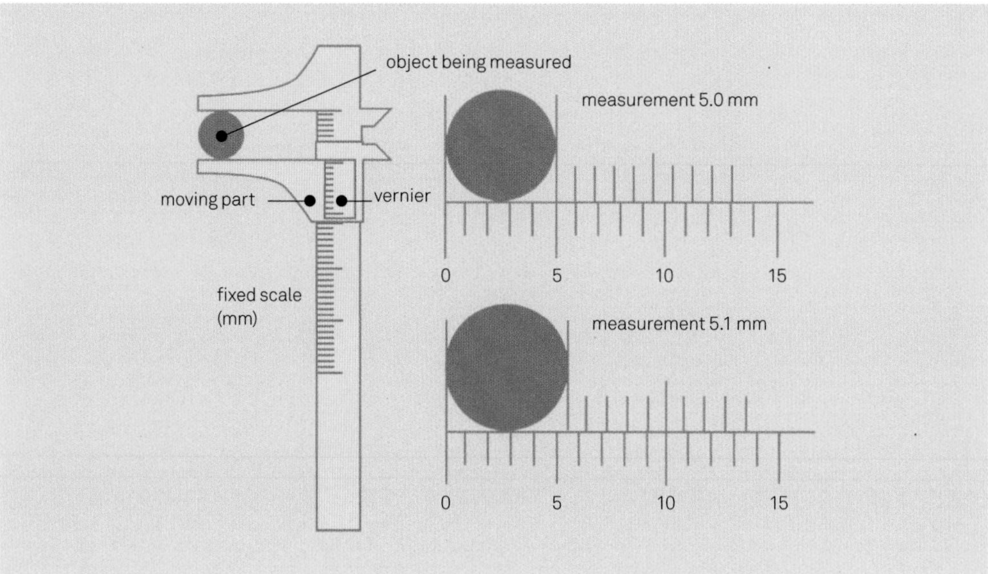

The vernier scale allows small measurements to be subdivided into even smaller measurements.

concept has been around for many thousands of years. Reference is made to an odometer made by Vitruvius around 25 B.C.E., and it is thought that Archimedes may have invented the first one.

pedometer

an instrument for estimating distance traveled on foot. The user estimates their average stride length, and the pedometer measures the number of strides by means of a component that moves with the jolt of the user's step. The distance traveled is equal to the average stride length multiplied by the number of strides. Thanks to fitness trackers, cell phones and the **Global Positioning System**, however, traditional pedometers have become mostly obsolete.

Global Positioning System (GPS)

a catch-all term for any of several highly precise satellite navigation systems that have gone from specialized applications to near-universal use in the 21st century. The original GPS system (also known as NavStar) has been developed and operated by the U.S. government since the 1970s, but similar systems have been launched by Russia, China and Europe. The system calculates a user's position on Earth by measuring the time delay in signals received from satellites whose orbits around the Earth are known with great precision, and can pin down a user's location, and record their movements, with an accuracy of a meter or less.

Geology

age of Earth

the oldest known Earth rocks date from around 3.8 to 3.9 billion years ago, but contain minerals that are 4.1–4.2 billion years old. These ages are established by **radio-isotope dating** and by studying meteorites believed to have formed at the same time as the planets.

The geochronological table. Pre-Cambrian time consists of three eras: the Hadean, the Archaean and the Proterozoic. Earth is believed to have been formed in the Hadean era 4,500 million years ago (m.y.a.) as a solid planet, but with no evidence of life.

Eon/era	Period	Epoch	M.Y.A.	Notes
Hadean			4,500	Earth forms as a solid planet. No evidence of life.
Archaean			4,000	Solid crust forms. Earliest single-celled life.
Proterozoic			2,500	Mountain ranges begin to form. Earliest multicellular life. Movement of tectonic plates slows to about the current rate.
Paleozoic	Cambrian		541	Metazoans (sponges and corals) and trilobites appear. Supercontinent Gondwana begins to break up.
	Ordovician		485	Earliest fish, but most life still invertebrate. No life out of water.
	Silurian		444	First sharks appear, along with the first plants on land.
	Devonian		419	Ammonites, amphibians and first air-breathing arthropods.
	Carboniferous		359	First flying insects appear, plant-life well established on land. Late in this period the first reptiles appear.
Mesozoic	Permian		299	Trilobites extinct. Supercontinent Pangaea formed.
	Triassic		252	First dinosaurs and earliest mammals appear on land.
	Jurassic		201	Birds evolve. Pangaea breaks up; the Atlantic forms.
	Cretaceous		145	First flowering plants. Dinosaurs and ammonites wiped out. Current contents emerge, but in different positions.
Cenozoic	Paleogene	Paleocene	66	Inland seas dry up. Ungulates, rodents and primates evolve.
		Eocene	56	Alpine-Himalayan and Rocky mountains begin to form.
		Oligocene	34	Grasslands expand at expense of forests. First apes appear.
	Neogene	Miocene	23	Higher primates evolve. Climate cools, Antarctica freezes.
		Pliocene	5	First appearance of genus *Homo*.
	Quatemary	Pleistocene	2.6	First appearance of *Homo sapiens*.
		Holocene	0.012	Earliest known civilization begins.

radio-isotope dating

radio-isotope dating relies on finding out how much an isotope has decayed over the years into its "daughter" isotope. For example, zircon crystals, which form in cooling magma, capture radioactive uranium-235, but not lead. Because there was no lead initially, and uranium-235 decays to lead-207, the amount of lead there now is an indication of the rock's age. The half-life of uranium-235 is 704 million years, so half the uranium-235 atoms originally present in a sample of zircon will be lead-207 after 704 million years. Other isotopes used for dating rock include rubidium-87 (half-life: 48,800 million years; daughter: strontium-87) and potassium-40 (half-life: 1,280 million years; daughter: argon-40).

carbon dating

radio-isotope dating using an isotope of carbon, carbon-14, whose "daughter" isotope is nitrogen-14. Carbon dating is useful because carbon-14's half-life is relatively short (in geological terms!)—just 5,730 years—so is good for dating items between 500 and 50,000 years old. As with all radio-isotope dating, care has to be taken to ensure that a sample has not been contaminated, for example by carbon-dioxide emissions from volcanic eruptions.

clinometer

an instrument for measuring the angle of inclination in rock **strata**. Movements of the Earth's crust often cause layers of rock that were originally laid down horizontally to slope at a significant angle.

The volcanic explosivity index (V.E.I.) measures the explosivity of an eruption, the volume of ash (tephra) and the height of the resulting ash cloud column. During human history no volcanic eruption has reached 8 on the V.E.I. scale, though this does not mean one of such magnitude has not occurred during the geologic past.

Volcanic explosivity index

	0	1	2	3	4	5	6	7	8
General description	non explosive	small	moderate	moderately large	large	very large			
Volume of tephra (m^3)		10^4	10^6	10^7	10^8	10^9	10^{10}	10^{11}	10^{12}
Cloud column height (km)*	<0.1	0.1-1	1-5	3-15	10-25	25			
Qualitative description	gentle, effusive	← explosive →		← cataclysmic, paroxsmal, collosal →					
			← severe, violent, terrific →						
Classification	← Strombolian →		← Plinian →						
	Hawaiian	← Vulcanian →			← ultra-Plinian →				
Total historic eruptions	487	623	3,176	733	119	19	5	2	0
1975-85 eruptions	70	124	125	49	7	1	0	0	

** Note: For VEI's 0-2 data are in km above crater, for VEI's 3-8, data are in km above sea level.*

volcanic explosivity index

a scale used to gauge the intensity of volcanic eruptions. It also includes a description of the eruption and a name for it. The number of eruptions in history is also recorded.

tectonic drift

the Earth's crust is made of a series of sections, called tectonic plates, which constantly move against one another, causing major natural features of the landscape, particularly mountain ranges. They typically move at up to about 2 inches (5 cm) a year. There are seven major tectonic plates, and a number of smaller ones.

strata

the individual layers of sedimentary rock, laid down over time. The layers of different types of rock are often visible, e.g., on cliff faces.

Richter scale

a measure of earthquake magnitude, named after American seismologist Charles Richter (1900–85). It is a logarithmic scale, beginning at 0, with each subsequent whole number representing ten times the magnitude of the previous one, and about 32 times the amount of energy released. It differs from the **Mercalli scale** in that it uses data recorded on seismographs, not observed effects of an earthquake on buildings and other structures. The earthquake in the Indian Ocean in December 2004 measured 9.3 on the Richter scale, and as such was one of the largest ever recorded.

Mercalli scale

a scale of the effects ("intensity") of earthquakes. Named after Italian seismologist Giuseppe Mercalli (1850–1914), who built on the work of physicists de Rossi and Forel to devise a scale indicating the observed effects of earthquakes on the ground —from 1 (effects only detected by seismographs) to 12 (total destruction). Further modified by American seismologists Harry Wood and Frank Neumann, the scale is now usually called the modified Mercalli scale.

moment magnitude scale

a successor scale to the **Richter scale** for measuring the magnitude of earthquakes. The scale was developed by Hiroo Kanamori, a Japanese seismologist who recognized that the

Mercalli scale

1 Not felt by people.
2 May be felt by people at top of buildings. Hanging objects begin to swing.
3 Vibration like passing small truck. Hanging objects set in motion.May not be recognized as earthquake.
4 Vibration like passing of heavy truck. Standing cars may rock, dishes rattle.
5 Felt outdoors. Liquids slop out of cups. Small objects may topple over.
6 Felt by all. People may be frightened. Pictures fall off walls. Glass may crack.
7 Difficult to stand. Weak chimneys break. Fall of plaster, tiles, cornices. Waves on ponds.
8 Steering of cars affected. Fall of some masonry walls. Severe structural damage to buildings. Changes in flow or temperature of springs.
9 Large-scale damage to buildings, dams, embankments. People panic, animals run.
10 Most buildings destroyed. Landslides, water thrown from rivers.
11 Destruction of roads, rails and underground services. Large cracks in ground.
12 Damage total. Large rock masses and water courses displaced. Lines of sight and level distorted. Objects thrown into the air.

Comparison between the Richter and Moment Magnitude (M.M.) scales

Earthquake	*Richter*	*M.M.*
New Madrid, MO, 1812	8.7	8.1
San Francisco, CA, 1906	8.3	7.7
Prince William, AK, 1964	8.4	9.2
Northridge, CA, 1994	6.4	6.7

Richter scale had a problem of "saturation" at high values, i.e., the higher up the scale, the less distinction between the magnitudes of earthquakes. The moment magnitude scale takes account of low-frequency seismic waves, which can often cause the most damage to large buildings.

sedimentation rate

the rate at which particles suspended in water or air settle. Natural sedimentation occurs at different rates with different materials and conditions, so measurements of sedimentation for particular materials can give an indication of how conditions have changed over time.

Zhubov scale

a scale for measuring ice coverage. Devised by a Soviet naval officer, N.N. Zhubov, the scale uses a unit called a ball. Clear water is 0 balls, 10 percent ice coverage is 1 ball, 20 percent is 2 balls, etc.

latitude

an imaginary line circling the globe parallel to the Equator. The Equator has a latitude of 0 degrees, and there are 90 degrees between the Equator and each of the Poles. The Tropic of Cancer lies at 23.5° north, and the Tropic of Capricorn at 23.5° south. Each degree represents about 69 miles (111 km) in distance at ground level.

longitude

an imaginary line, also known as a meridian, along the surface of the Earth between the North and South Poles. Zero degrees longitude is the Greenwich or prime meridian, passing through Greenwich in London, England. The line of longitude farthest from the Greenwich meridian is 180° east or west, which is also essentially the location of the International Date Line. With **latitude**, longitude forms part of a grid reference system that can identify the position of any place on Earth.

North and South Poles

the Poles are the points at maximum distance from the Equator in the Arctic and Antarctic respectively, and the points of convergence of lines of **longitude**. Earth has a magnetic field whose poles roughly coincide with the North and South Poles, so adjustment must be made to compass readings to get true bearings—the farther north you go the more important this is.

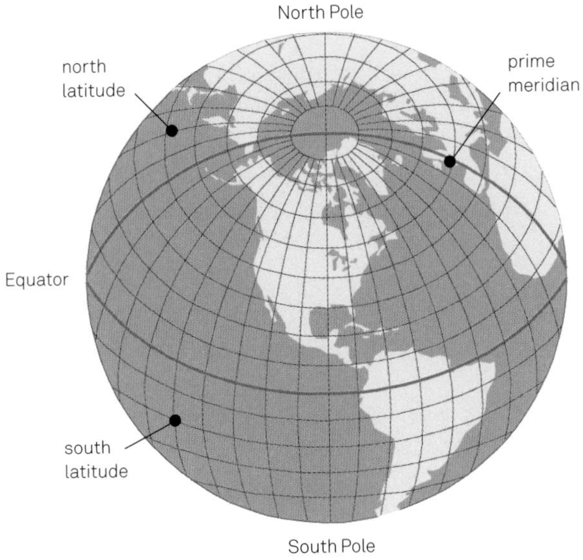

The imaginary lines crisscrossing the Earth's surface provide a grid reference system to identify the location of any place.

Magnetic North is about 625 miles (1,000 km) from True North, and is constantly moving at 6–25 miles (10–40 km) a year. The polarity of the Earth's magnetic field has completely reversed many times in the past.

strike and dip

two words to describe the angle of inclination to the horizontal of layers of rock or a fault. Strike indicates the direction of a line that is formed where the inclined plane and the horizontal intersect. It is the angle between this line and True North. Dip is the tangent to the surface of Earth at the point where the dipping surface cuts it.

gravimeter

a gravimeter is an instrument that measures Earth's gravitational field at different points on the surface, allowing comparison to be made between the different points. It is particularly useful in prospecting for oil and minerals.

magnetic declination (magnetic deviation)

the difference between the directions of True and Magnetic North (*see* **North and South Poles**). Because of the constant change in location of Magnetic North, magnetic declination is also constantly changing.

Land area

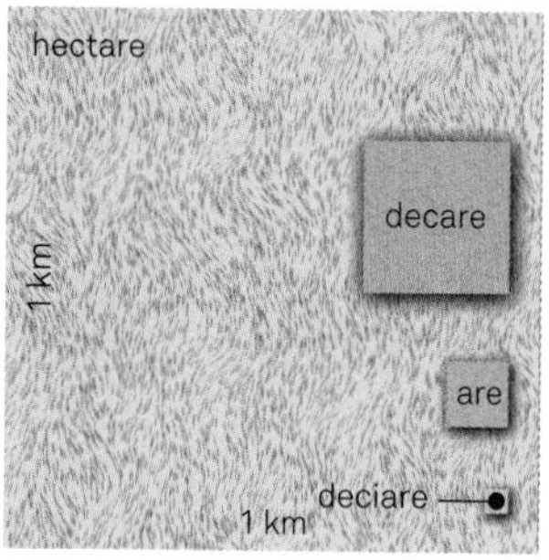

The hectare is 1 square kilometer. One hectare consists of 10 decares, 100 ares or 1,000 deciares.

hectare (ha)

a metric unit of land area equal to one square kilometer or approximately 2.471 acres. Derived via French from the word *are* and the Greek *hekaton*, it is now in more common usage in the U.S. than elsewhere in the world.

square meter (m^2)

the fundamental metric unit of area. A square meter is the area enclosed by a square whose sides are each 1 meter in length. Just as the meter is the basic SI unit of length, so the square meter is the basic SI unit of area. The Conférence Générale des Poids et Mesures (C.P.G.M.) recommends that all areas be measured in terms of square meters rather than, say, hectares or square kilometers. A square meter, of course, does not have to be square: an area 4 meters long and 25 cm wide is still a square meter.

acre (ac)

an imperial unit of land area equivalent to 4,840 square yards or 4046.8564 square meters. Historically the acre was a much less precise unit of measurement, being traditionally defined as the area of land a man and an ox could plow in one day. For this reason, it was originally not a square measurement but a long strip, this being a more time-efficient area to plow as it required few turnings of the plow itself. It was defined as the current area of 4,840 square yards by the British Weights and Measures Act of 1878. The U.S. survey acre is fractionally smaller than the international acre, measuring 4,046.8726 square meters.

rood (ro)

an imperial unit of land area equivalent to a quarter of an acre, or approximately 0.1012 hectares. Originally, a rood was an area of land equivalent to 40 rods long by one rod wide. The rood has now all but fallen out of usage.

hide
a non-specific unit of land area. Formerly used across Britain, a hide was deemed to be the area of land that a family and its dependants needed to support themselves. Of course, the amount of land required by a family depended not least on the size of the family, so the hide was generally taken to be anything between 60 and 120 acres. Over time the hide also came to be a measure of tax liability. The Anglo-Saxon Chronicle records that in C.E. 1008 the king (Aethelred II) demanded that every 300 hides should produce one warship and every 8 hides a helmet and a coat of chain mail.

hundred 👁
historically in Britain, a subdivision of a county or shire which came to be used as a unit of land area.

riding
an ancient British subdivision of land. Derived from the Old English word *trithing*, which was itself a derivation of the Old Norse *thrithjungr* meaning "third part," the riding was traditionally one of three subdivisions of the county. The term still exists in Yorkshire, which is divided into East Riding, North Riding and West Riding.

Alfred the Great is believed to have been the first king of England to divide shires into hundreds.

county
a subdivision of a country. Originally, a county was that area of Britain under the jurisdiction of a count or earl. Counties are now territorial subdivisions which form a principal unit of local administration.

section
a unit of land area principally used in U.S. and Canadian land surveying. A section is normally equivalent to 1 square mile, or 640 acres, although sections are sometimes slightly smaller or larger to compensate for the curvature of the earth.

township
a unit of land area principally used in U.S. and Canadian land surveying. A township is made up of 36 sections and is equivalent to 36 square miles or 23,040 acres. The township was first defined as a unit of land area by an act of 1785 which stated that its two sides must run north–south, and that the other two sides must be at right angles, i.e., that the township should be square or rectangular. This did not apply if natural features such as rivers made it impractical, or if the township would then cover a Native American reservation.

Townships consist of 36 sections, and sections are equivalent to 1 square mile, with some exceptions to allow for the Earth's curvature.

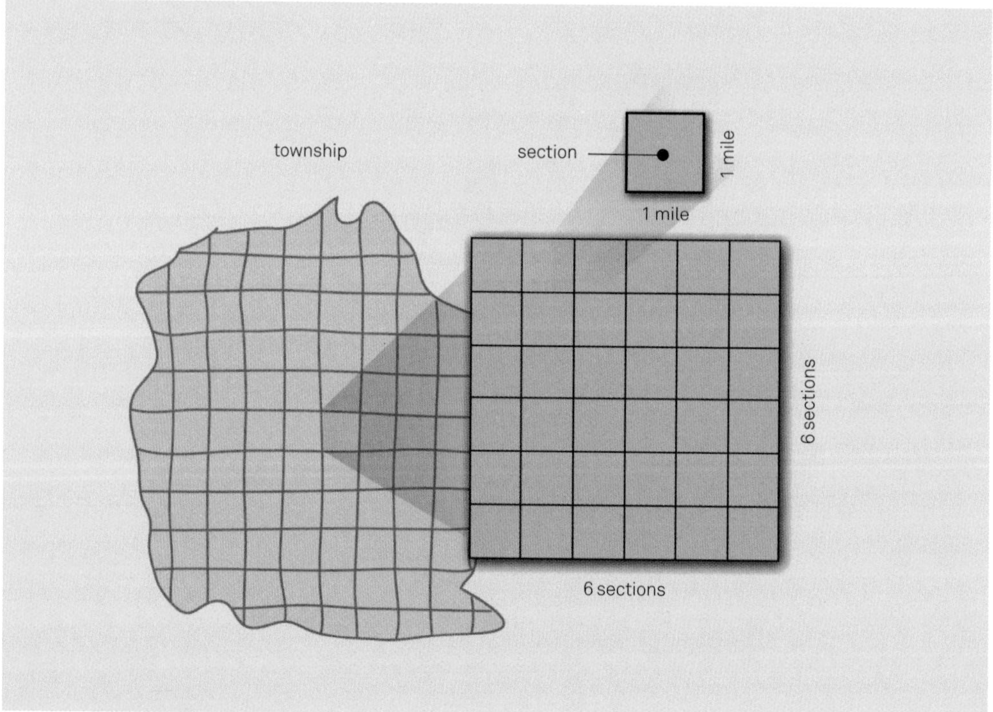

square mile (sq mi)
an imperial unit of land area. A square mile is the area enclosed by a square whose sides are each 1 mile in length, but it need not be square itself: an area 4 miles long by a quarter of a mile wide is still a square mile. It is equivalent to 640 acres or approximately 2.5 square kilometers.

square inch (sq in)
an imperial unit of area. A square inch is the area enclosed by a square whose sides are each 1 inch in length, but it need not be square itself. It is equivalent to 6.4516 square centimeters.

hacienda
a unit of land measurement deriving from Spanish territories in South America. A *hacienda* was traditionally an area of land awarded to Spanish nobles in Mexico, Argentina and other areas of South America. It is officially calculated as being 89.6 square kilometers, but in fact the area represented by the original *haciendas* varied enormously.

actus quadratus
a Roman unit of land area. The *actus quadratus* was the basic Roman unit of land measurement, and was a square of 120 by 120 Roman feet, or *pes*. This made it 14,400 square *pes*, which approximates to 13,500 square feet today.

arpent
the principal unit of land area in France between the 16th and 18th centuries. It was defined as being 100 square *perches*, but this was not a fixed measurement as the *perche* varied according to the region of France. Thus, the *arpent de Paris* was equivalent to approximately 3,420 square meters—this was the most common arpent; the *arpent commune* was equivalent to approximately 4,220 square meters; and the *arpent d'ordonnance* (also known as the *arpent des eaux et forêts* or the *grand arpent*) was equivalent to approximately 5,100 square meters. The *arpent* also exists in Canada, as a result of the French influence there. The Canadian *arpent* derives from the *arpent de Paris*, and so is also equivalent to approximately 3,420 square meters.

ching
a Chinese unit of land area, equivalent to approximately 13.3 square meters or approximately 143 square feet.

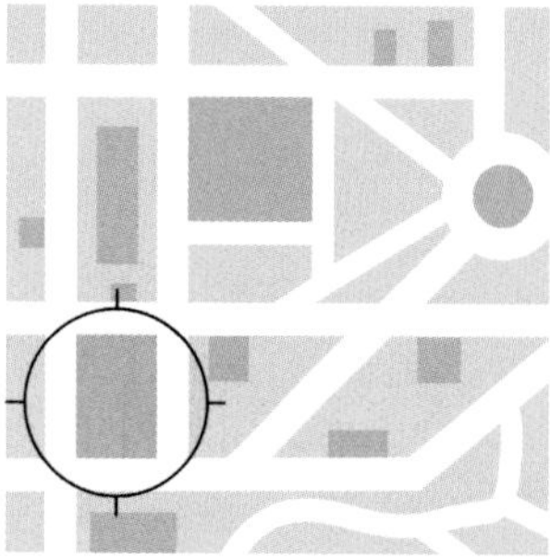

Where a block is rectangular, rather than square, the distances between parallel streets are referred to as "long blocks" and "short blocks."

ch'ing

a Chinese unit of land area. The *ch'ing* is much larger than the *ching*, being approximately 1 hectare or 2.47 acres.

mu

a Chinese unit of land area. Throughout China's history, the definition of a *mu* has varied wildly, having been as small as 192 square meters during the early Zhou dynasty and as large as 840 square meters during the Yuan dynasty. In 1959, the *mu* was standardized as a metric unit equivalent to 666 $^2/3$ square meters.

feddan

a unit of land area used in Egypt and the Sudan. It was historically used throughout North Africa and the Middle East, and is approximately equal to 4,200 square meters.

morgen

a traditional unit of land area used historically throughout northern Europe. Deriving from the German word for morning, a morgen was the amount of land a yoke of oxen could plow in a morning. As this is of necessity an imprecise measure, the morgen became accepted as different areas in different countries.

chomer

an ancient Hebrew unit both of land area and capacity. Translated variously in modern English versions of the Bible as "homer" and "measure," as a capacity the *chomer* was equivalent to about 230 liters, and as land area was deemed to be the amount of land a *chomer* of seed would plant—approximately 2.4 hectares or 6 acres.

rai

a Thai unit of land area. The word *rai* means "field" (an upland field rather than a paddy field). It is an ancient unit of measurement, which is now accepted to be equal to 1,600 square meters or about 0.4 acres.

block

a non-specific unit of land area, principally in the U.S. and Canada. Most North American cities form regular street grids. A block is the area of land enclosed by four intersecting streets or, colloquially, the distance from one street to the next parallel street. Distances between streets vary a great deal from city to city, and are generally anywhere between 80 meters and 160 meters. In some cities, such as New York, the streets running in one direction are closer together than the streets running perpendicular to them, giving rise to the terms "long block" and "short block."

zip code/postal code

alphanumeric systems of dividing land areas into postal regions. ZIP is an acronym for Zone Improvement Plan, and consists of five digits. An extended zip code, called ZIP+4, adds a further four digits to the standard zip code and allows an address to be pinpointed precisely. The first number in the standard zip code represents a particular group of states; the second number represents a region within that group; and the final three numbers define a more specific area within that region. The postal code is a sequence of letters and numbers. The first letter or letters represent a region; the following one or two numbers represent a district within that region; and the final group of letters and numbers represents a more specific area. An individual postal code can represent a street, part of a street or even a single building.

continent

any one of the world's expanses of land. The word "continent" derives from the Latin *terra continens*, meaning "continuing tract of land"; as such it is a non-specific unit of measurement, and the number of continents into which the world is divided is not universally agreed. However, it is generally accepted that there are up to seven: Europe, Asia, Africa, North America, South America, Australia and Antarctica.

The world is generally divided into seven continents, though Asia and Europe form one continuous tract of land and Australia is sometimes viewed as part of the Asian land mass.

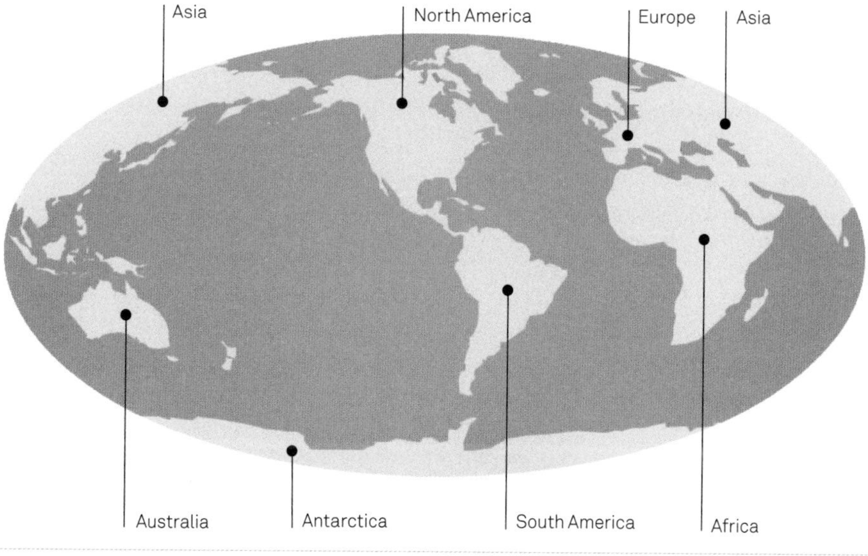

Commerce

payload
the capacity of a commercial aircraft for carrying revenue load (that is, its freight, mail, passengers and baggage). The total available capacity of an aircraft, measured in tonnes, is known as the payload capacity; the amount of revenue load carried, again in tonnes, is the payload carried.

ton-mile (tonne-kilometer)
a unit used in commercial aviation for calculating the costs of moving revenue loads, relating to moving 1 ton of load 1 mile (or 1 tonne of load 1 kilometer).

tonnage
in shipping, the measurement of capacity of a vessel for purposes of registration, assessing tolls, etc. Despite the nomenclature, tonnage is a measure of volume rather than weight, and usually refers to gross tonnage: the capacity available for cargo, stores, fuel, passengers and crew. Until the International Convention on Tonnage Measurement of Ships in 1969, tonnage was normally expressed in gross registered tons (1 grt = 100 cubic feet), but nowadays gross tonnage (gt) is defined in terms of cubic meters using the formula:

$gt = K_1V$
where $K_1 = 0.2 + 0.02 \log_{10}V$
and V = space in cubic meters

The net registered tonnage (nrt) is calculated by a complicated formula involving the draft, passenger-carrying capacity and a coefficient Kc, while a further, even more complex formula is used to determine compensated gross tonnage (cgt), which also takes into account the size and type of vessel.

Deadweight (dwt), mercifully, is more easily calculated. It is the weight in U.K. tons or tonnes of the cargo, stores, fuel, passengers and crew of the vessel carrying its maximum

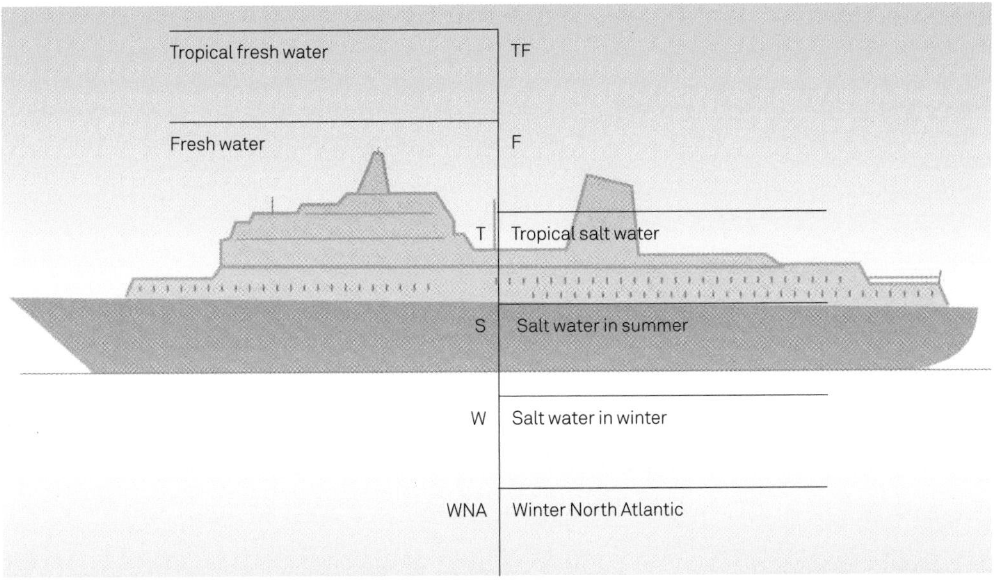

summer load. Lightweight tonnage (lwt) is the weight of the vessel expressed in terms of the weight of water it displaces in tons or tonnes.

Plimsoll line (loadline)
a mark on the side of a ship showing the maximum permissible draft (distance of keel below waterline) in various seasons and waters. Although officially referred to as the loadline, it is also known widely as the Plimsoll line or mark after British politician Samuel Plimsoll (1824–98), who campaigned tirelessly for stricter regulation of sea-going vessels.

twenty-foot equivalent
a unit used in the transport of containerized cargo. In shipping, this is normally quoted in 20-foot equivalent (TEU) or in 40-foot equivalent (FEU), which denotes the length of the container, though other length containers are used for some cargoes. The standard width for these containers is 8 feet (2.4384 meters), and until recently the height was also 8 feet, but these days 40-foot containers are more usually 9 feet 6 inches (2.8956 meters) in height.

tun
a unit of liquid measurement used particularly for wines, spirits and beer in the English-speaking world. It represented the largest size of cask in common usage, and while sizes varied

Any boat, ship, barge or floating wharf floats at different levels depending on the warmth, consistency or both, of the water. When a ship is fully loaded, the loadline appropriate to the waters in which it is traveling and the season should not be submerged.

somewhat, the standard tun contained 200 U.K. gallons (or 252 U.S. gallons—also known as "wine gallons") or 953.88 liters. A tun could be divided into 2 pipes, 4 hogsheads, or 6 tierces. Nowadays the beer, wines and spirits wholesale trade is conducted predominantly in liters or hectoliters.

barrel 👁

a unit of volumetric measurement used historically for trading in liquids such as wines and spirits, and for certain dry goods. Today the barrel is recognized internationally only as a measure of petroleum and 1 barrel is 42 U.S. gallons or 158.987 liters. However, in the English-speaking world the barrel is still used for measurement of a wide variety of goods, with a confusing number of different values. Beer barrels vary in size between the U.S. and U.K. (1 U.S. barrel is 31 U.S. gallons or 117.35 liters, while 1 U.K. barrel is 26 U.K. gallons or 163.66 liters) and even between U.S. states. For dry goods, fruit and vegetables the U.S. barrel is 7,056 cubic inches—except for cranberries, where it is 5,826 cubic inches! As if things weren't complex enough, the barrel is also sometimes used as a unit of weight. The Canadian barrel of cement is 350 lb, whereas in the U.S. it is 280 lb for masonry cement , but 376 lb for Portland cement.

kilderkin

a variable unit of volumetric measurement for liquid and dry goods, equal to half a barrel in U.S. usage, and consequently dependent on the size of barrel. In the U.K. the term loosely describes a cask of about 16 to 18 U.K. gallons (between 73 and 82 liters).

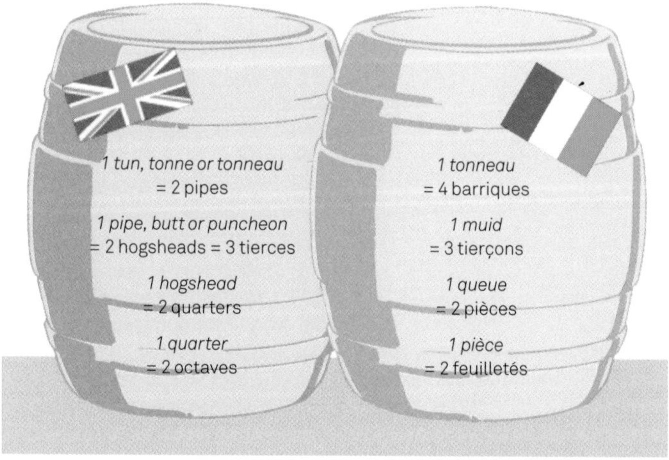

Terms and sizes of containers for beers, wines and spirits. The actual volumes of these containers vary widely according to region, and there are also many different-sized containers specific to particular drinks or areas, such as the butt, aum, leaguer and stück.

firkin
a unit of volumetric measurement for liquid and dry goods, equal to half a kilderkin. In the U.S., the firkin is also used as a measurement of weight, 56 lb (25.401 kg), probably the approximate or average weight of goods contained in a firkin.

bottle sizes 👁

bottle sizes for wine range from the "half bottle" at 375 ml to the Nebuchadnezzar at 15 liters. The larger sizes are usually reserved for sparkling wines, particularly champagne. Wines are normally retailed in 75 cl (or sometimes liter) bottles, spirits in 70 cl (sometimes 75 cl or 1 liter), and by the case of 12 bottles. Cheap wine is sometimes sold in liter bottles. Beer and cider commonly retail in bottles of 275, 330, and 500 ml and 1, 1.5, 2 and 3 liter sizes, and in the U.K. draft beer and cider are sold in pubs in pint or half pint measures.

amphora
a container for wine or oil used in ancient Greece and Rome. As with the barrel, the name of the container could also be used for a specific size—around 25 liters in the late Roman period—and the two should not be confused, as the container came in many sizes. The amphora measure was divided into three *modii*, or two *urnae*, and was used for both liquid and dry goods.

The sizes and names of bottles traditionally used for champagne; in Bordeaux, however, the jeroboam is 5 or 6 bottles, and the methuselah is known as an impériale, while in the U.K. the jeroboam is 6 bottles, and the rehoboam 8.

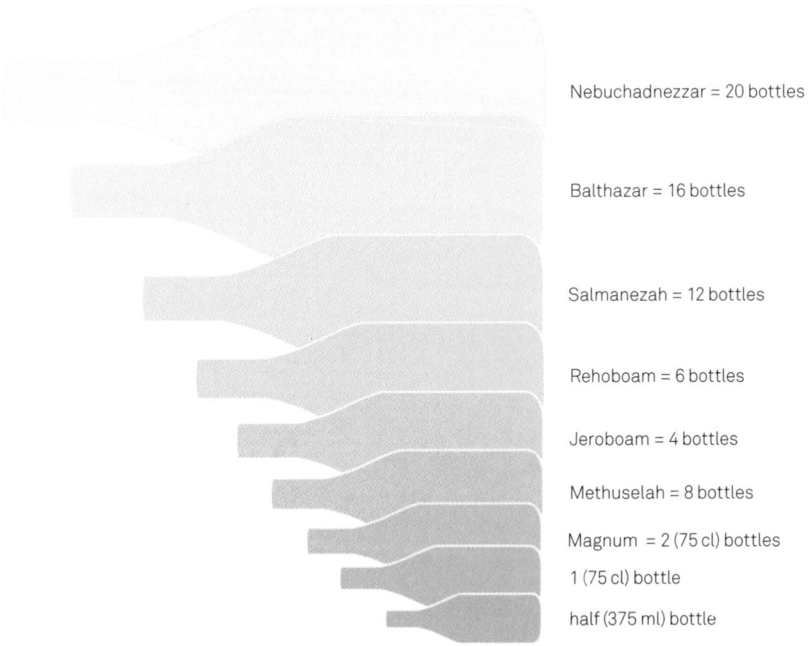

hectoliter
one hundred liters in the metric system—a measure used widely in wholesale trading of liquids. It has largely replaced the traditional units such as the barrel, tun, etc., in most trades except petroleum, although there is some resistance to its adoption in some parts of the world, including the U.S., which still use imperial measures.

keg
a container, usually a small cask, for various goods, varying in size and definition accordingly. In the wine trade a keg traditionally measured 12 U.S. gallons (about 45.52 liters), or half a barrel of beer, with correspondingly different values in the U.K. and U.S. In the fishing industry a keg represented 60 herring. As a measure of nails it was equal to 100 lb (45.359 kg).

bushel
a unit of volume with several different values, though now largely obsolete. In the U.S. it is a unit of dry measurement, also known as a Winchester bushel, equivalent to 4 pecks, or about 1.2445 cubic feet (35.239 liters). In the U.K. the bushel is a liquid measure, equivalent to 8 U.K. gallons (36.369 liters). In commercial trading throughout the English-speaking world, agricultural produce such as grain was measured by weight in bushels of various sizes according to the commodity and the country. The bushel was later standardized as 60 lb (27.216 kg), and known as the international corn bushel, but has now been superseded by metric measures.

peck
a unit of volume, a subdivision of the bushel, used primarily for measurement of dry goods such as grains and fruit. A peck is the equivalent of quarter of a bushel, or 2 gallons, which in the U.S. equates to 8.80975 liters and in the U.K. to 9.09225 liters.

sheaf
a traditional (and approximate) unit for measuring grains such as wheat and barley still on the stalk. A sheaf is a bunch of stalks roughly 30–36 inches (75–90 cm) in circumference.

cord
a unit of volume used in forestry to measure stacked roundwood, still widely used in the U.S. It is defined as the volume of a stack of logs 8 feet long, 4 feet wide and 4 feet high, and thus is equal

to 128 cubic feet (3.625 cubic meters). The cord foot, is ⅛ of a cord, or 16 cubic feet (0.4531 cubic meters), while a rick is equivalent to ⅓ of a cord.

board foot
a traditional unit of volume for timber, also known as foot board measure (fbm), board measure, or super foot. One board foot describes the volume of a board 1 foot wide, 1 foot long and 1 inch thick, therefore equivalent to 1⁄12 of a cubic foot (0.00283 cubic meters).

hoppus foot
a traditional unit of volume for timber used in U.K. forestry, named after Edward Hoppus, who devised it to measure the volume of wood in a log of length L and circumference or girth, G (in feet):

Usable timber (hoppus ft) = $L\,(G/4)^2$.

The hoppus foot is equivalent to 1.273 cubic feet (0.0361 cubic meters). Related units still used in some parts of the world are the hoppus ton, equivalent to 50 hoppus feet (1.8027 cubic meters), and the hoppus board foot, equivalent to 1⁄12 of a hoppus foot.

standard
a traditional unit of volume for stacked timber used in forestry. There were three such standards in common use: the St. Petersburg or Petrograd standard (165 cubic feet), the Göteborg or Gothenburg standard (180 cubic feet), and the English standard (270 cubic feet).

cunit
a traditional unit of volume for timber used in forestry. It represents the volume of usable wood, i.e., excluding the bark and gaps between logs, and one cunit is the equivalent of 100 cubic feet (2.8317 cubic meters) of solid wood.

ring
a unit of quantity used by coopers for boards and staves in barrel-making. The term comes from the metal ring used in transportation to hold the staves together in bundles of 240, and thus represents a quantity of 240 staves or boards. It was divided into four shocks of 60 boards.

Medicine

apgar score

a means of measuring the health of a baby immediately after it is delivered. Ranging between 0 and 10, the score is derived from observations of five criteria: heart rate, respiratory effort, muscle tone, reflex irritability and skin color. The baby is given a score between 0 and 2 for each criterion at 1 and 5 minutes after birth. Each child, therefore, has two apgar scores. Below 3 is considered critical and the baby may require medical attention; a score of above 7 is considered normal.

The apgar score, named for Virginia Apgar, who devised it in 1952.

Sign	Score = 0	Score = 1	Score = 2
Heart rate	Absent	Below 100	Above 100
Respiratory effort	Absent	Weak, irregular or gasping	Good, crying
muscle tone	Flaccid	Some flexion of extremeties	well flexed, or active movements of extremities
Reflex irratability	No response	Grimace/weak cry	Good cry
skin colour	Blue all over, or pale	Acrocyanosis	Pink all over

body mass index (B.M.I.)

a ratio of a person's weight in kilograms to the square of their height in meters, and a loose indication of health. The body mass index is a means of calculating whether an individual is underweight, overweight or the correct weight for their height. Generally speaking, an index of less than 18.5 indicates that a person is underweight, whereas an index of above 25 indicates that a person is overweight. These calculations are only an approximate guide to health, however: recommended B.M.I.s alter with age, and do not take into account the amount of body fat—a sportsperson with a high percentage of muscle may have the same B.M.I. as an obese person, but will not necessarily be overweight.

blood pressure (mmHg)
a measure of the pressure of the blood in the large arteries. Blood pressure is measured in terms of two numbers: the maximum or systolic pressure (normally 100–135 mmHg in healthy humans) minimum or diastolic pressure (normally 50–90 mmHg). Thus, a blood pressure of 120/80 would indicate a systolic pressure of 120 mmHg and a diastolic pressure of 80 mmHg.

sphygmomanometer
a means of measuring **blood pressure**. An inflatable cuff is placed around the upper arm, and the cuff inflated. The air is then slowly let out of the cuff, reducing the pressure on the arm. When the cuff is full of air, it measures the systolic blood pressure. When the air has been released, it measures the diastolic blood pressure.

blood count
a means of measuring the number of corpuscles in a given volume of blood. Blood contains a large number of different particles, the quantities of which can be a good indication of health.

clotting time
the time taken for blood to clot, a measure of the efficiency of clotting, and by extension of the general health of the blood. The clotting time of the blood of a healthy human can be anywhere between 5 and 15 minutes.

blood group
a means of describing the characteristics of human blood. The most common descriptions of blood groups are those of the ABO system. This defines the blood group (A, B or O) according to what type of antigen (A or B) the red blood cells carry on their surface, and what type of antibodies they produce. Certain blood groups are compatible with others, while others are incompatible.

electrocardiogram (E.C.G.)
a record of a person's heart rate and other cardiovascular functions, produced by means of electrocardiograph (a recording of the electric voltage in the heart as a continuous strip graph). In popular culture, an E.C.G. with a flat line indicates that death has occurred. In fact, a "flatline"—technically known as a systole—indicates cardiac arrest with a very bad prognosis, but not necessarily death.

resting heart rate

the number of contractions of the heart in 1 minute when a person is resting. The average resting heart rate is 70. A resting heart rate of less than 60 is a symptom of bradycardia, although it is not necessarily a cause for concern unless it is accompanied by other symptoms of the condition. Very fit people, such as sports men and women, often have a resting heart rate of below 60.
A resting heart rate of above 100 is a symptom of a condition called tachycardia.

metabolic rate

the number of calories the human body burns. The basal metabolic rate indicates the caloriec consumption requires to maintain basic bodily functions at rest, and varies according to age, weight, height, diet, fitness and other factors. Metabolic rate increases as a result of exercise, and can be affected by certain conditions such as an overactive thyroid.

electroencephalogram (E.E.G.)

a visual representation of the electrical activity of the brain. An E.E.G. is a neurophysiologic exploration used to assess brain damage, epilepsy, etc., recorded by means of electroencephalography, which involves attaching electrodes to the scalp. E.E.G.s are presented as visual lines either on paper or on an oscilloscope.

Electroencephalograms record the electrical activity of the brain over time. A flat line where electrical activity = 0 would indicate no brain activity.

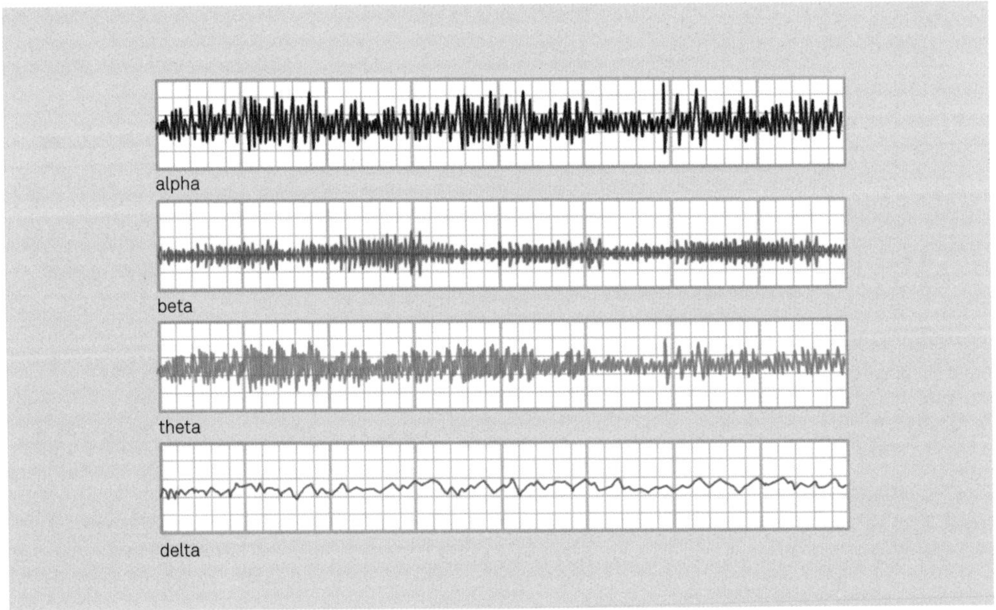

brain waves

a measure of the electrical activity of the brain, as represented by an **electroencephalogram**. E.E.G.s measure four different types of brain wave—alpha, beta, delta and theta—each falling within different ranges of frequency. The type of brain wave emitted by the brain alters according to both the state of consciousness and the age of the subject.

I.Q.

a measure of a person's cognitive abilities. I.Q. stands for intelligence quotient. A person's I.Q. is measured according to a series of standardized tests, and is normalized so that the average I.Q. of sets of people of certain ages is 100. An I.Q. of more than 100, therefore, indicates that the person's intelligence quotient is above average compared to people of the same age; an I.Q. of less than 100 indicates that it is below average.

twenty-twenty vision

a measure of vision acuity which indicates that a person's vision is perfectly normal, as measured on a chart devised by Dutch ophthalmologist Hermann Snellen (1834–1908). In order to score 20/20, an individual must be able to read the 20th line of Snellen's chart at a distance of 20 feet.

lung capacity

a measurement of the volume of air that can be absorbed by the lungs during the deepest possible inhalation. Also known as vital capacity, it may be approximated in cubic centimeters by multiplying the surface area of the body by 2,500.

peak flow

a measure of the maximum rate of air breathed out during forced expiration. Peak flow is measured using a peak-flow meter, a short calibrated tube with a mouthpiece.

respirometer

an instrument used to measure the rate of respiration of an organism. Respirometers measure the intake of oxygen and the output of carbon dioxide.

Breathalyzer

an instrument used to measure the amount of alcohol in a person's breath. As blood passes through the lungs, the air causes a proportion of any alcohol in the bloodstream to evaporate. The alcohol content of the breath, therefore, is directly proportional to the alcohol content of the blood, which is why Breathalyzers give a reliable indication of the alcohol level of the blood. The ratio of blood alcohol to breath alcohol is generally around 2,100:1, though this can change from person to person.

Glasgow coma scale (G.C.S.)

a means of measuring a patient's response to head trauma. The Glasgow coma scale, or score, is calculated by awarding a score to each of three observations: eye response, verbal response and motor response. Each observation is given a score between 1 and 5, and the total of the three scores produces the G.C.S. A score of 3 (the lowest score) indicates that a person is in a deep coma; a score of 15 (the highest score) indicates that a person is fully awake.

sun protection factor (S.P.F.)

a measure of the effectiveness of a sunscreen against the effects of ultraviolet type B radiation which causes sunburn. A sunscreen of S.P.F. 10 means that the user can stay in the sun for 10 times longer than they would be able to without it, although this figure is affected by other factors such as the user's skin type and the strength of the sun. S.P.F. ratings do not indicate the sunscreen's ability to filter ultraviolet type A radiation, which can also have damaging effects on the skin.

burns, degrees of

a measure of burn severity. There are commonly three degrees of burn: first degree, second degree and third degree. The symptoms of a first-degree burn are redness and soreness; a second-degree burn also has some level of blistering; and a third-degree burn includes some charring of the skin. Burns that affect the tissue beneath the skin are sometimes referred to as fourth-degree burns.

computed tomography

a process used to produce a detailed image of a cross-section of the body. Computed tomography uses a series of X-rays taken around a particular axis of rotation. The images thus produced are known commonly known as C.A.T. scans.

ultrasound scanning

a means of producing a two- or three-dimensional image of internal organs using high-frequency sound waves.

Ultrasound is very good at imaging muscle and soft tissue, and has the advantage that it can produce real-time, moving images on a screen; it cannot, however, penetrate bones.

magnetic resonance imaging (M.R.I.)
a means of producing a two-dimensional image of internal organs using magnetism and sound waves. M.R.I. scans resemble X-ray images, although they are somewhat more detailed and have the advantage of not exposing the body to potentially harmful X-ray radiation.

positron emission tomography (P.E.T.)
a means of producing three-dimensional, color images of processes occurring in the body, such as metabolic functions. P.E.T. scans involve the introduction of short-lived radioactive substances into the body, whose radioactive isotopes can be tracked.

radiography
the process of creating images on photographic film using X-rays. X-rays have the ability to pass through solid objects, but the strength of the resulting ray is dependent on the density of the object it has passed through. Radiographs can therefore display a two-dimensional image of the internal structure of an organism.

X-rays pass less readily through bone than through soft tissue, which is why radiography can produce useful images of the skeleton.

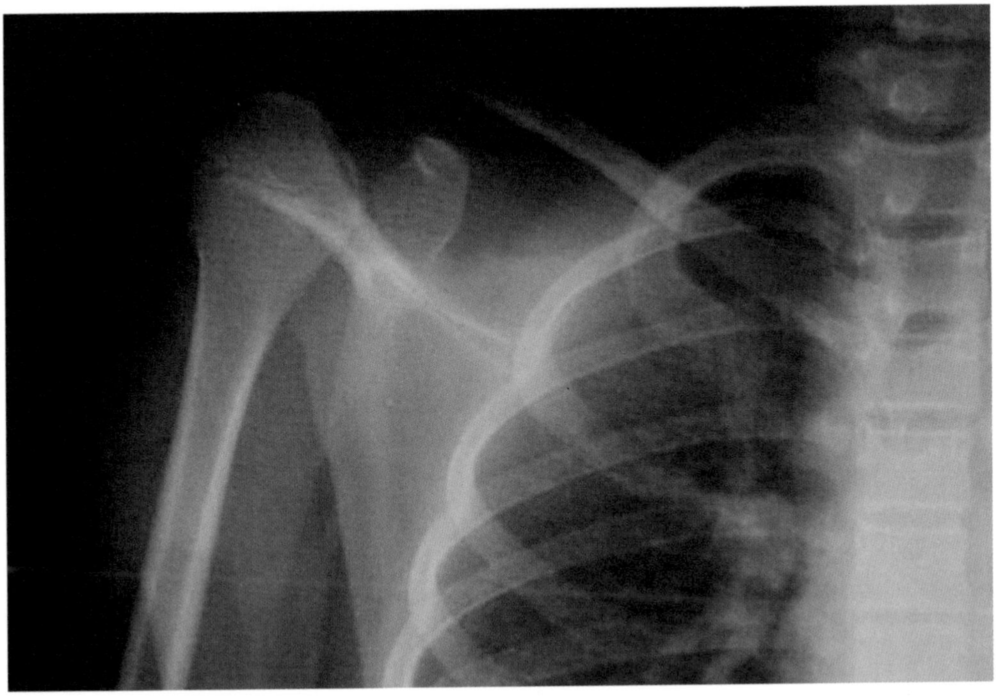

posology
the branch of medicine concerned with dosage. The word posology derives via French from the Greek *posos*, meaning "how much."

L.D.$_{50}$
a measure of toxicity. L.D.$_{50}$ stands for "lethal dose 50 percent," and is defined as the dosage of a substance that kills 50 percent of a trial population. L.D.$_{50}$ is generally expressed in terms of mass of the substance per body mass—e.g., milligrams per kilogram. It is gradually being phased out as a measure of toxicity because it of necessity refers to animals rather than humans.

mouse unit (M.U.)
a unit of toxicity. A mouse unit is the amount of a substance that causes death in 50 percent of mice—i.e., it is the L.D.50 for mice. So, the size of a mouse unit varies according to the toxin.

demography
the study of size, structure and distribution of populations. The study of demography includes factors such as incidence of disease, birth rates, death rates, fertility rates, infant mortality rates, life expectancy and reproduction rates. Demographic data can come from a wide range of sources, including birth records, death records and census information.

birth and death rates
the number of births and deaths per population of 1,000 people per year; these are more properly known as the crude birth rate and crude death rate, and can be misleading. A more meaningful statistic of mortality is provided by a death rate according to age.

life expectancy
a measure of the average remaining lifetime of an individual in a given population. In populations with a high infant-mortality rate, therefore, life expectancy at birth is considerably different from life expectancy at, say, age five. Colloquially, however, the term life expectancy refers to life expectancy at birth.

Meteorology

altitude
in meteorology the height of an object above the surface of Earth, above mean sea level or above a constant-pressure surface.

altimeter
an instrument for measuring the altitude of an object. A pressure altimeter measures barometric pressure and compares it to barometric pressure at sea level; a radio altimeter measures the time taken for a radio signal to travel from a transmitter on the Earth's surface to an object and back to a receiver; and the **Global Positioning System** (G.P.S.) measures the time taken for radio signals to travel between satellites and a receiver.

lapse rate
the rate of fall of an atmospheric variable (normally temperature) with increase in height.

Weather maps primarily show the position of areas of high or low pressure by means of isobars, and the position and type of weather fronts.

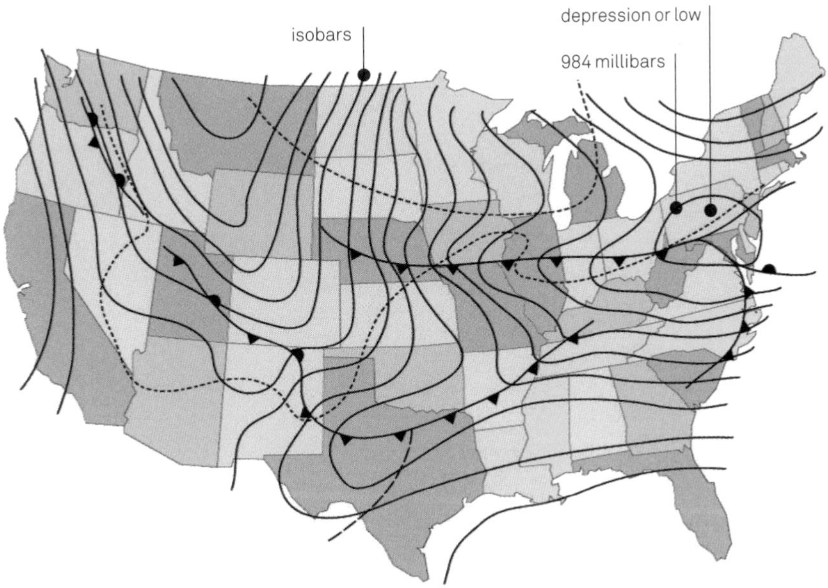

The atmospheric pressure acting on the surface of a bath of mercury is sufficient to support a column of mercury. The space at the top of the calibrated tube is known as the Torricellian vacuum.

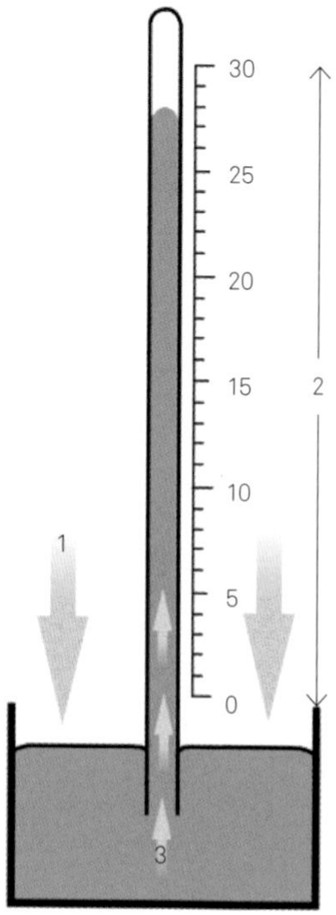

1 Air pressure presses down on the mercury, forcing it up the evacuated glass tube.
2 Scale measures height of mercury.
3 Covering keeps mercury from wspilling, but allows air pressure to influence mercury.

bar (b)
a unit of atmospheric pressure equal to 1.01972 kg of force per square centimeter (about 14.50376 lb force per square inch); this is equivalent to a little more than the average pressure of the Earth's atmosphere (1.01325 bar). For practical meteorological purposes, the more common unit is the millibar (mb). A related unit, the barye (ba), is the C.G.S. unit of pressure and equals one microbar. The words derive from the Greek *barys* (weight), from which we also get the terms baromil, barometer, barograph, etc.

millimeter of mercury (mm Hg)
a unit of pressure, equivalent to the pressure of a column of mercury one millimeter high at the Earth's surface. The unit comes from the use of mercury barometers, where the pressure can be read as the height of the column of mercury. In meteorology both millimeters and inches of mercury have now been replaced by the bar and millibar. In medicine, however, blood pressure is still measured in mm Hg.

barometer
an instrument for measuring atmospheric pressure. A mercury barometer consists of a vertical glass tube containing mercury, closed at the top and immersed in a bath of mercury at the bottom; pressure is read by the height of the column of mercury in this tube. Less accurate, but more portable and compact, is the aneroid barometer, which consists of a vacuum chamber of thin corrugated metal in the form of a bellows, with one end fixed and the other attached to a mechanism for converting the movement of the bellows caused by variations in atmospheric pressure to a pointer on a scale.

thermobarograph
an instrument for recording pressure and temperature, a combined thermograph (recording thermometer) and barograph (recording barometer).

isobar
a line on a weather map connecting places having the same atmospheric pressure at a given time.

isotherm
a line on a weather map connecting places having the same temperature at a given time.

atmosphere 👁

the sphere of gases that surrounds the surface of Earth. The Earth's atmosphere extends to an altitude of about 2,500 km, and can be divided into different layers: the heterosphere, comprising the exo-sphere and the thermosphere; and the homosphere, comprising the mesosphere, stratosphere and troposphere.

wind speed

the speed of movement of air across the Earth's surface, measured in meters per second, or traditionally in miles per hour or knots.

Beaufort wind scale

an empirical scale for estimating wind speeds based on observable effects of the wind. It was originally devised by Admiral Sir Francis Beaufort (1774–1857) and was based on the effect of the wind on waves at sea. Later it was developed as a scale for use on land, with corresponding observable effects on land. The internationally recognized scale of 0 to 12 was extended to 17 by the U.S. Weather Bureau in 1955, but the additional force numbers 13 to 17 are generally considered impractical and unnecessary.

wind chill factor

a measurement of the cooling effect of wind speed combined with low temperature on the surface of the human body, refined in 2001 to give the wind chill temperature index (W.C.T.I.) according to the formula, wind chill = $13.12 + 0.6215T - 11.37(V^{0.16}) + 0.3965T(V^{0.16})$, where T is the temperature in °C, and V is the wind speed in km/h.

Coriolis force

a fictitious force used to explain the movement of objects in a rotating system, specifically in meteorology to account for the deviation from the expected movement of bodies due to the rotation of Earth. It can be seen in the apparent drift eastward of winds blowing north from the Equator, and arises from the comparative difference of speeds of rotation at different latitudes. The Coriolis effect is named after French mathematician Gaspard Gustave de Coriolis (1792–1843).

Fujita tornado scale

an empirical scale for measuring the wind speed of tornadoes based on observable damage caused by them, more correctly called the Fujita–Pearson Scale. It is similar to the Beaufort scale in its use of force numbers, ranging from F0 to F5, to denote increasing wind speeds.

Altitudes of the different levels of atmosphere are:

1 exosphere 700–2,500 km
2 thermosphere 85–700 km
3 mesosphere 50–85 km
4 stratosphere about 12–50 km
5 troposphere up to about 12 km

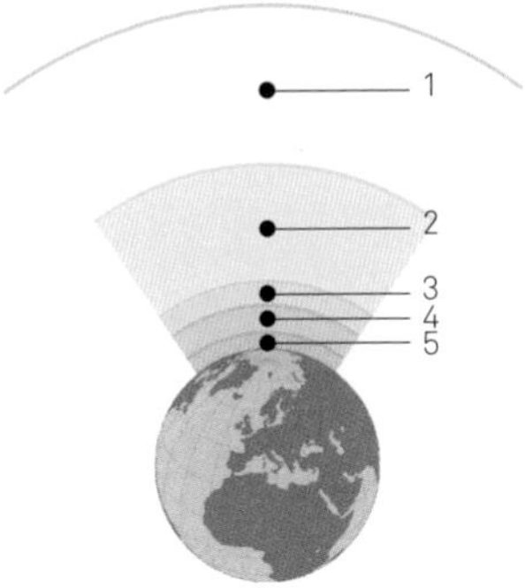

At the time of the spring tides, high tides are higher and low tides lower than average; at the time of the neap tides, the difference between high and low tides is smaller than average.

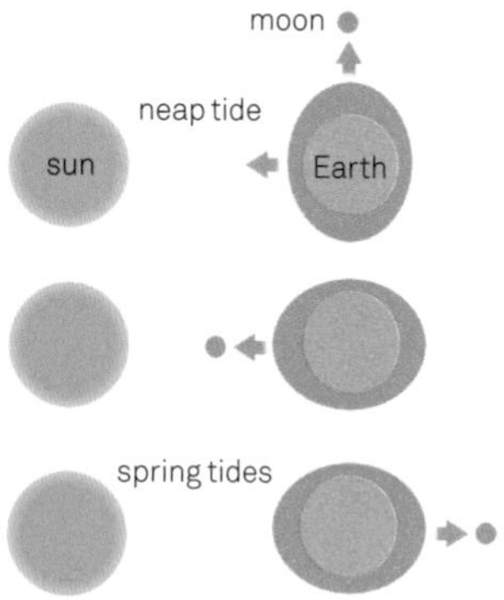

tropical cyclone intensity scale
a scale for describing the severity of revolving storms (including cyclones, hurricanes and typhoons) that originate over the tropical oceans according to their average wind speeds at standard anemometer level (10 meters). The internationally agreed scale of intensity is:

1. tropical depression, with winds up to 17 meters per second
2. tropical storm, with winds of 18–32 meters per second
3. severe tropical cyclone, hurricane or typhoon, with winds of 33 meters per second or higher.

Saffir–Simpson hurricane scale
a scale for describing the severity of hurricanes based on observable damage caused by them.

drought severity scale
a scale for measuring the severity of droughts in specific locations at specific times, taking into account the precipitation deficit and the abnormality of the weather. In 1965, American meteorologist W.C. Palmer devised a scale known as the Palmer Drought Severity Index (P.D.S.I.), which was calculated from precipitation and temperature figures, and the available water content of the soil.

tides (neap and spring)
in meteorological terms, tides are specifically the ocean tides on Earth caused mainly by the gravitational force of the moon, and to some extent by that of the sun. This attraction causes a "bulge" in the water of the oceans facing toward the moon, and a corresponding bulge on the opposite side of Earth, which are the areas of high tides. When the sun's gravity reinforces that of the moon (at the time of full and new moons), there are stronger tides known as spring tides. When the sun's gravity is acting at 90° to the moon's (at the time of half moons), there are weaker tides known as neap tides.

humidity
the amount of moisture in the air. It is usually (and more correctly) expressed as relative humidity, the ratio of vapor pressure in moist air to the saturation vapor pressure with respect to water at the same temperature, given as a percentage. The relative humidity of a given sample of air is greatly affected by temperature, and conversely the relative humidity affects the apparent temperature. This relationship can be seen in a heat index chart (also called a comfort chart), similar to a wind chill chart, which gives the perceived temperature at various temperatures and relative humidities.

dewpoint
the temperature at which a given sample of moist air becomes saturated and deposits dew when in contact with the ground or a cooler surface, or condenses into water droplets above ground (when it is occasionally called the cloud point).

hygrometer
an instrument for measuring atmospheric humidity. There are several types of hygrometer, including the wet and dry bulb hygrometer (also known as a psychrometer) which consists of two thermometers, one with its bulb surrounded by a damp wick dipping into water, which gives a reading affected by the cooling effect of the relative humidity of the ambient air, the other giving a "true" reading of the temperature. A calculation can then be made of relative humidity by comparison of the two readings.

rainfall
the amount of moisture falling on the Earth's surface at a given point in a given period, usually expressed in centimeters or inches. Meteorological usage has firmly established the term "rainfall," when "precipitation" would provide a more accurate description, as this term includes snow and hail.

pluviometer
an instrument for measuring rainfall, also known as a rain gauge.

hyetograph
a chart or diagram showing the time and amount of precipitation in a given place. The data for hyetographs can be taken from a continuously recording rain gauge.

sonde
a device carried in a balloon, satellite or rocket to send back data such as measurements of temperature, pressure and humidity. A radiosonde transmits this data by radio to a ground receiver, and can send details of the successive levels of the atmosphere as it ascends to the stratosphere.

inversion
a reversal of the expected decrease in temperature with altitude in the atmosphere. An inversion layer is a layer of air that is warmer than the layer below it. Inversions are not uncommon, and occur on clear nights and in anticyclones.

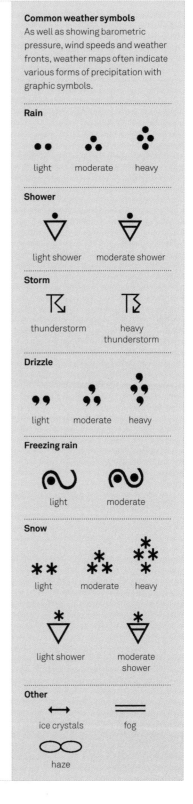

Common weather symbols
As well as showing barometric pressure, wind speeds and weather fronts, weather maps often indicate various forms of precipitation with graphic symbols.

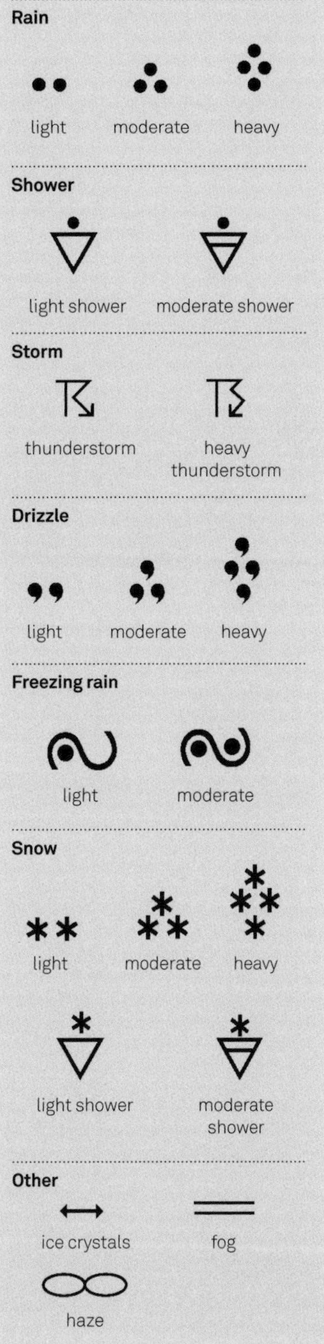

visibility

the maximum distance at which a sufficiently large dark object can be seen against the sky at the horizon in normal daylight, or a moderately intense light can be seen at night. Normally, visibility readings are made at several points around the horizon circle and averaged out to give a single reading for that location.

transmissometer (transmittance meter)

an instrument for measuring visibility, or more precisely the transmission or extinction coefficient of the atmosphere to determine visibility. It is also known as a telephotometer or hazemeter.

cloud base

the lowest altitude at which the air contains a measurable quantity of cloud particles in a given cloud or cloud layer, also known as base of cloud cover. The height of the cloud base above local terrain is known as the cloud height, and the vertical distance from cloud base to the cloud top is known as the thickness or depth of cloud.

cloud types

clouds are classified according to shape, and the height they occur, and were given their Latin names by meteorologist, Luke Howard, in 1803. *Cirrus* means curl of hair, *stratus* means layer, *cumulus* means heap. There are 10 cloud types, each starting at one of three levels: high clouds, above 20,000 feet (6,000 m); mid-level clouds, at 6,500 to 20,000 feet (2,000-6,000m); and low clouds, below 6,500 feet (2,000m).

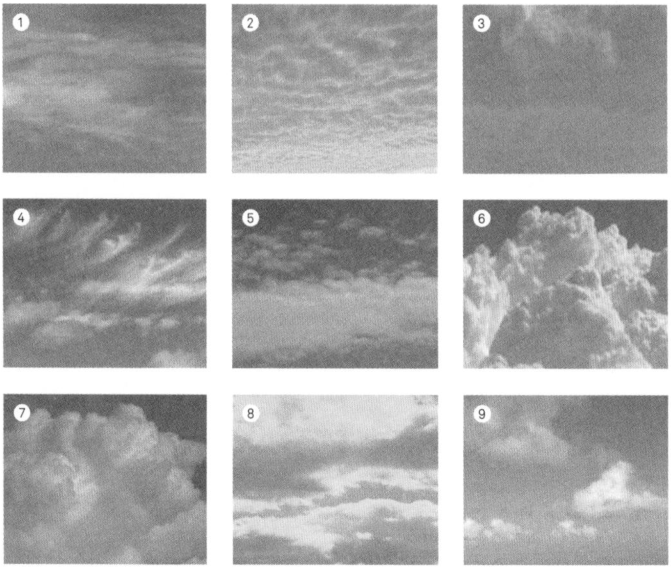

1 *Cirrus*: detached wispy white clouds of ice crystals.
2 *Cirrocumulus*: small round clouds of ice crystals and water droplets, often regular ripple-like formations.
3 *Cirrostratus*: continuous white veil of cloud formed of ice crystals.
4 *Altostratus*: gray/blue thick sheet of cloud containing water droplets.
5 *Altocumulus*: gray or white round clouds, often ripple-like formations.
6 *Cumulonimbus*: vertically developing and towering clouds, dark at the base, with anvil shaped white tops.
7 *Cumulus*: vertically developing, billowy white "cotton-wool" cloud, often with gray base.
8 *Stratus*: sheets/patches of gray, shapeless cloud, often starting as fog.
9 *Stratocumulus*: sheets/patches of round gray cloud.

Minerals and metals

streak
the color of a material in powdered form, and therefore the color that a "streak" drawn with the material across a harder substance will take. Often a useful way of identifying materials that otherwise look similar to the eye.

permeability
the ability of a substance to allow liquid or gas to flow through. It depends on the size of the substance's pores, and the extent to which they are connected together.

assay ton (A.T.)
a measure of the amount of precious metal contained by an ore, equal to the number of troy ounces of pure metal that would be obtained from a ton of ore. In conversion to metric units, the assay ton suffers from the difference between the long (British) ton of 2,240 lb and the short (U.S.) ton of 2,000 lb; the former giving a ratio of 1 gram of metal to 32.7 kg of ore, while the latter is 1 gram of metal to only 29.2 kg ore.

ore grade
the proportion of the desired metal(s) in an ore, usually given as a percentage of the total weight for common metals, or as parts per thousand (ppt) or parts per million (ppm). Precious metals and others (such as uranium) with low concentrations will often be given in grams per tonne.

cut-off
the lowest ore grade from which it is worth extracting a metal. The level of cut-off is an economic factor in determining whether a particular vein is worth mining. Cut-offs vary enormously from one material to another—the economic cut-off for iron ore is around 55–60 percent, while gold can be mined profitably at less than a gram per tonne in open-cast workings.

The average abundance in the Earth's crust and cut-off grades for various metals.

	Element	Average abundance (ppm)	Minimum ore grade (ppm)	Factor
	Gold	.004	0.5	125
	Molybdenum	2	500	250
	Tin	2	500	250
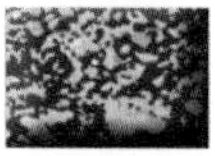	Lead	12	15000	1250
	Copper	55	5500	100
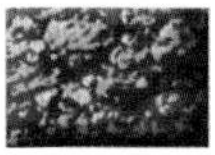	Zinc	70	5000	700

trace

a very low concentration of a mineral, sometimes defined as anything less than 1 part per million. Sometimes this will be the main desired substance (for precious metals, for example), but it can also be a convenient (or inconvenient) impurity in a vein.

reserves

the amount of ore, fossil fuel, or other valuable substance contained within a particular deposit or **vein**. Normally quoted as a number of tonnes (or barrels, etc.) of the ore on the assumption that the **cut-off** remains the same throughout the deposit. For metals, it is sometimes quoted as a number of tonnes of ore with a specified average grade. Mining and oil companies also state their total reserves, simply the sum of the reserves in all their claimed deposits.

sieve number

a measure of the opening size in a mesh sieve, and therefore of the largest particle that can pass through it. The sieve (or mesh) number is the number of holes per unit distance.

slate sizes

roofing slates come in a variety of standard sizes; as with many other traditional materials, these sizes are given names rather than the more modern habit of merely referring to the size itself. A "wide" slate is 2 inches wider than standard; a "small" one 2 inches shorter—thus a wide Duchess is 24″ × 14″ and a small Lady 14″ × 8″.

phi scale

a scale of particle size based on average diameter, intended primarily for sediments, sand and other small objects. A 0ϕ object is (approximately) 1 mm in diameter; the diameter is halved for every step in the scale (so a 2ϕ object is 0.25 mm across). Negative numbers are used for larger objects—a 1-inch diameter object is about –4.7ϕ.

A.S.T.M. grain size index

a measure of the size of the "grains" (or crystals) making up the internal structure of a metal or other substance. The grain size index is N where the number of grains per square inch at 100 times magnification is $2^{(N-1)}$. (So 1 for 1 grain per square inch, 2 for 2, 3 for 4 grains per square inch, etc.)

metal fatigue

a process by which metal objects suffer a gradual loss of strength, and eventually break as a result of fluctuating stresses that always remain below the tensile strength of the object. Continually changing stresses cause slow changes in the structure of the material, allowing a crack to form at a point of stress concentration and then slowly extend until it reaches the **Griffith crack length** and breaks completely.

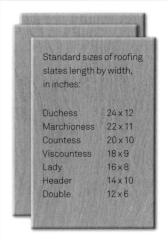

Standard sizes of roofing slates length by width, in inches:

Duchess	24 x 12
Marchioness	22 x 11
Countess	20 x 10
Viscountess	18 x 9
Lady	16 x 8
Header	14 x 10
Double	12 x 6

The traditional names for roofing slates have been said to be falling into disuse for many years now, but can still be found on the Web sites of numerous firms selling traditional slates.

Phi units*	Size	Wentworth size class	Sediment/rock name
-8	256mm	Boulders	Sediment gravel
		Cobbles	
-6	64mm	Pebbles	Rock, rudites (conglomerates, breccias)
-2	4mm	Granules	
-1	2mm	Very coarse sand	Sediment: sand
0	1mm	Coarse sand	
1	1/2mm	Medium sand	
2	1/4mm	Fine sand	Rocks: sandstones (arenites, wackes)
3	1/8mm	Very fine sand	
4	1/16mm	Silt	Sediment: mud Rocks: lutites (mudrocks)
8	1/256mm	Clay	

*Udden-Wentworth scale

fatigue limit

the stress below which a material does not suffer any fatigue damage, regardless of the number of times the stress is repeated. Many materials (e.g., aluminum) do not have a fatigue limit, and will eventually fail completely under repeated loadings with even the smallest stress; the magnitude of the stress only changes the number of fatigue cycles that the material can survive. Fatigue strength is a related but different measure, being the stress at which the material will only fail after at least a specified number of cycles.

ductility

a combined measure of the ease with which a material can be drawn into wires, rods, sheets, etc., and the ability of the material to withstand such treatment.

malleability

the ease with which a material can be hammered (or otherwise compressed) into shape, and how much forming of that type it can accept.

karat (kt), carat (ct.) 👁

a measure of the purity of gold, being the number of 24 equal parts by weight that are gold. Thus 24 karat gold is pure (as far as possible), 18 karat is 75 percent gold, and so on. "Colored" white and red gold can never reach a purity of more than around

Hallmarks have been used in Britain and France since at least the 14th century. Modern British usage has a crown and a fineness number—before 1975, the karat purity was used for gold. The shape of the "shield" around the number indicates the metal. The lion (see below and right) indicates 92.5 percent silver, while the leopard's head (see right) is one of several marks used to show where the metal's purity was checked. Other countries use different systems (the U.S. has no official system at all), often much more complex. Bullion bars for trading and transport are normally "fine silver," meaning 999 fine.

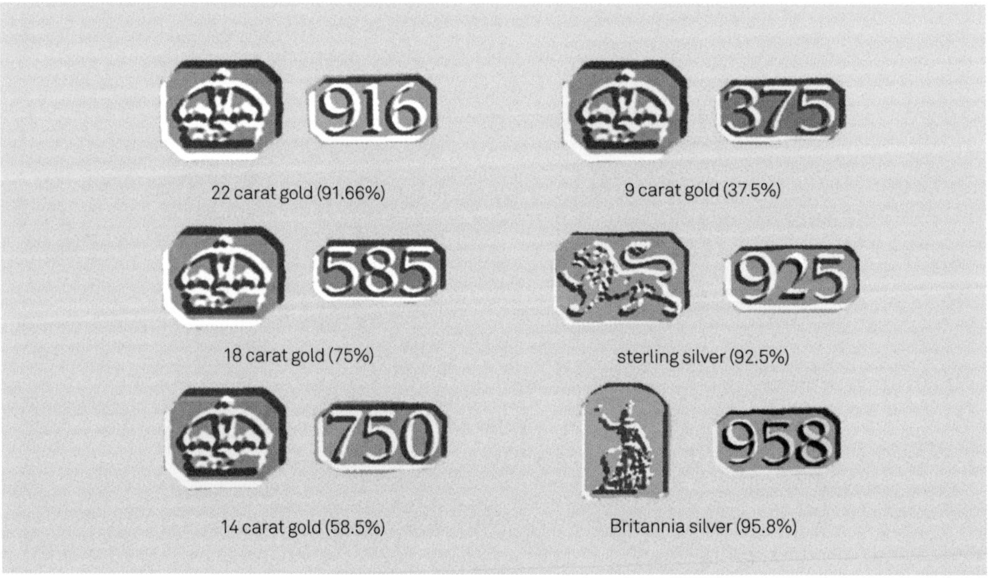

22 carat gold (91.66%)

9 carat gold (37.5%)

18 carat gold (75%)

sterling silver (92.5%)

14 carat gold (58.5%)

Britannia silver (95.8%)

18 karat, since the color is produced by the impurities (copper in red gold, and nickel and platinum in white). Karat (metal purity) and carat (a unit of mass used for gemstones) are used differently in the U.S., though the root is the same, and the two are identical in many places; in the U.K., carat is used for both; in Germany, *Karat*. As a unit of mass, a value of 200 mg was officially adopted for the carat (also known as the metric carat) in 1907.

fineness

a measure of the purity of precious metals, equal to one part in a thousand. Expressed as a number between one and 1,000, so 18 karat gold is 750 fine.

silver grades 👁

a variety of names exist for standard purities of silver. "Coin silver" is 90 percent silver (the other 10 percent is usually copper), and used for making commemorative coins in various countries. "Mexican silver" is usually 95 percent silver and 5 percent copper. The two main grades used for jewelery and decorative items (and commemorative coins, in many countries) are "Sterling silver" (92.5 percent) and "Britannia silver" (95.8 percent).

gauge (of sheet metal and wire)

a measure of the thickness of a wire or sheet of metal. In traditional American and British systems in inches, higher numbers indicate thinner material. Negative numbers are not used, with the gauge thicker than 1 being 0, the next either 00 or 2/0, the next 000 or 3/0, and so on. There is, however, no fixed standard for the conversion from inches to gauge number. Metric wire gauges are much simpler, with the number being simply 10 times the thickness in millimeters and therefore increasing for thicker wires.

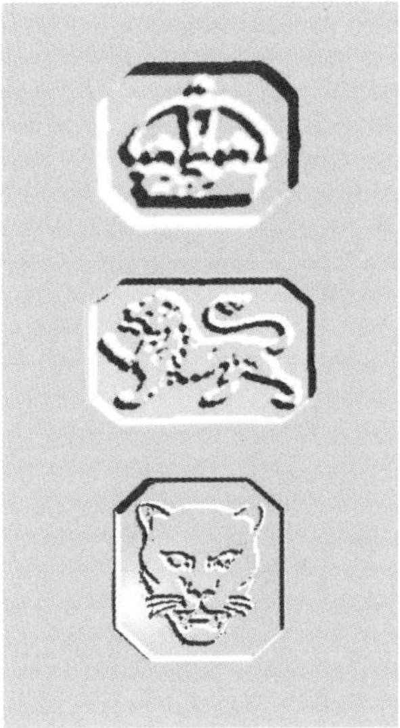

Time and calendar

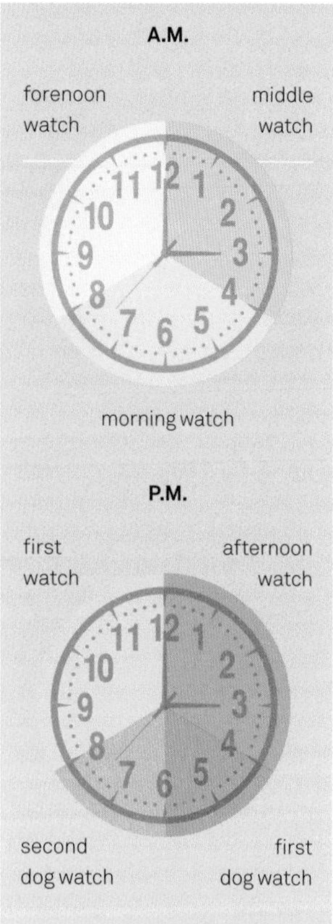

In horology the term ship's bell refers to a clock which strikes according to a system akin to that used on board ship where a bell is struck manually up to eight times to denote the four-hour "watches" or the periods of duty.

Planck time

a very small unit of time—probably the shortest unit of time possible within the present laws of physics. It is defined as the time taken by a photon to travel 1 **Planck length** at the speed of light, and is equal to about 1.351×10^{-43} second.

second (s; sec)

the base unit of time in the SI system. Originally defined as $^1/86{,}400$ of a mean solar day, then refined to specify $1/86{,}400$ of the mean solar day January 1, 1900, the internationally accepted definition was fixed in 1967 in terms of the frequency of radiation corresponding to the transition between the two hyperfine levels of the ground state of the caesium-133 atom, the duration of 9,192,631,770 periods of this particular radiation being exactly 1 second. From the second, other units of time, such as the minute and hour, can be derived.

minute (min)

a unit of time equal to 60 seconds. Before the adoption of the second as the base unit of the SI system, it was defined in terms of being $1/60$ of an hour, or $1/1440$ of a day. The **sidereal** day is divided in this way.

bell (ship's time) 👁

a unit of time, a division of the **watch**, used especially on board ships: each four-hour watch is divided into eight bells of 30 minutes. At sea, a bell is struck every half hour to mark the division of the watch, and is sounded the same number of times as the number of bells that have occurred since the beginning of that watch.

hour (h; hr)

a unit of time equal to 3,600 seconds, or 60 minutes. Before the adoption of the second as the base unit of the SI system, the hour was thought of, loosely speaking, as $1/24$ of a day, and the **sidereal** hour is in fact 1/24 of a sidereal day. The idea of a 24-hour day stems from the ancient division of the daylight into 12 periods (making summer hours longer than winter hours!), and the subsequent division of the nighttime also into 12. With the advent of mechanical clocks, the obvious next step was to have 24 equal hours to the day.

day (d; da)
a unit of time traditionally defined as the period taken for one rotation of Earth on its axis in relation to the sun; in everyday usage a period of 24 hours. This definition, however, is not as accurate as it may seem, and more specific ones have evolved. To astronomers, Earth completes one rotation in relation to distant stars in 23 hours 56 minutes 4.09054 seconds (the extra approximately 3 minutes 56 seconds is apparent to us on Earth only because of our orbit around the sun), and they refer to this period as the **sidereal** day. Here on Earth, however, the day is dictated by our relationship with the sun, and civil time, as opposed to sidereal time, is reckoned accordingly; the average period occurring between two successive noons (when the sun crosses the meridian) is known as the mean solar day. Our present definition of the **second** is independent of the behavior of the planets, so that days calculated as 86,400 seconds (60 × 60 × 24) are not in step with the mean solar day—particularly as this is increasing very gradually day by day—and occasional leap seconds have to be inserted to make up for the anomaly.

week
a unit of time equal to 7 days. It has been used for thousands of years, despite its awkward relationship with the 365-day year, probably because of religious and astrological associations.

lunar month
the period of time between two successive passages of the moon through opposition or conjunction, for example the period between two new moons. It is more properly known as the synodic month, and occasionally called a lunation. It is equal to 29.53059 days.

calendar month 👁
a division of the year according to a particular calendar, most usually the **Gregorian calendar**. Because the year is irregularly divided into 12 by the Gregorian calendar, the length of a calendar month can vary from 28 to 31 days. Other calendars, such as the

Jewish calendar months *no. of days*	**Gregorian calendar months** *no. of days*	**Islamic lunar months** *no. of days*	**Hindu lunar months** *no. of days*
Tishri 30	January 31	Muharram 30	Chaitra 30
Marcheshvan 29 or 30	February 28 or 29	Safar 29	Vaisakha 31
Kislev 29 or 30	March 31	Rabi'a I 30	Jyaistha 31
Tebet 29	April 30	Rabi'a II 29	Asadha 31
Shebat 30	May 31	Jumada I 30	Sravana 31
Adar 29 or 30	June 30	Jumada II 29	Bhadrapada 31
Nisan 30	July 31	Rajab 30	Asvina 30
Iyar 29	August 31	Sha'ban 29	Karttika 30
Sivan 30	September 30	Ramadan 30	Margasirsa 30
Tammuz 29	October 31	Shawwal 29	Pausa 30
Ab 30	November 31	Dhul-Qa'da 30	Magha 30
Elui 29	December 31	Dhul-Hijia 29 or 30	Phalguna 30

Note: Ve-Adar (29) is the Jewish intercalary month that features every 3rd, 6th, 8th, 11th, 14th, 17th and 19th year

The Jewish, Gregorian, Islamic and Hindu calendars compared. All the major calendars in use today have some form of intercalary unit to compensate for the awkwardness of the 365.242199 days in a tropical year.

Jewish and Chinese, have variable-length calendar months and years, whereas the Islamic calendar uses lunar months.

trimester

a unit of time equal to ¼ of a year, also known as a quarter. In medicine, the term is used to denote the three 14-week stages of the human gestation period, whereas in schools or colleges it means an academic term of about 14 weeks. It derives from the Latin, meaning three months.

semester

a unit of time equal to half a year, or 6 months. The term is principally used in schools and colleges to mean half an academic year, and thus can vary from about 15 to 21 weeks.

common year

a year in the Jewish calendar, which may take different lengths ranging from 353 to 385 days depending on the addition of a leap-days and a 13th intercalary month of 29 days called Ve-Adar.

sidereal year

an astronomical unit of time, the period taken for the Sun to return to the same exact location relative to the distant stars.

Sidereal time is reckoned by the movement of Earth in relation to distant stars, as opposed to civil time, which is measured relative to the sun; in sidereal terms Earth makes about 366.242 rotations during each revolution around the sun, So, despite the sidereal day being shorter than the mean solar day, the sidereal year is equal to 365.25636 days, a little longer than the **tropical year.**

year (a; y; yr) 👁

in general use, a period of time equal to 365 or 366 days, roughly the period taken for Earth to complete one revolution around the sun. A more precise definition is provided by the term **tropical (or solar) year.** As Earth actually takes about 365.242 days to make its way around the sun, calendars using whole numbers of days soon get wildly out of step with the true situation. Thus, the notion of a 365-day year with a leap year every fourth year was introduced by the **Julian calendar,** and further refined in the **Gregorian calendar**. In calendars of other cultures, a year can be anything from about 350 to 385 days.

tropical year (solar year)

the exact period of time taken for Earth to complete one revolution around the sun. The tropical year is equal to 31,556,925.9747 seconds, or about 365.242199 days.

The Hindu solar (zodiac) calendar. The astrological signs of the zodiac originated in Mesopotamia, but rapidly spread throughout Europe and Asia. The astrological year is divided into 12 months, each assigned its own astrological sign derived from the constellation in which the sun appears.

1 Maysha (Aries) the Ram
2 Vrushabha (Taurus) the Bull
3 Mithuna (Gemini) the Twins
4 Karka (Cancer) the Crab
5 Simha (Leo) the Lion
6 Kanya (Virgo) the Maiden
7 Tula (Libra) the Scales
8 Vrushchika (Scorpio) the Scorpion
9 Dhanu (Saggitarius) the Bow
10 Makar (Capricorn) the Goat
11 Kumbha (Aquarius) the Pot
12 Meena (Pisces) the Fish

leap year

a year in the **Gregorian calendar** in which an extra intercalary day has been added to account for the fraction of a day in the tropical year, making it a period of 366 days. Leap years normally occur every fourth year in the Gregorian calendar, except when the year is divisible by 100, unless it is also divisible by 400. Calendars in other cultures, e.g., the Jewish, Islamic and Chinese calendars, also have leap years which include intercalary days or months.

decade

a period of 10 years. The term decade was also used in the French Republican calendar for a period of 10 days in an attempt to decimalize the calendar.

generation

an approximate unit of time, the interval between the birth of a parent and the birth of its child. Necessarily open to various interpretations, dependent on culture, location, and historical period, it can be anything from about 20 to 35 years.

century

a period of 100 years. Because of the absence of the year 0, the centuries are somewhat confusingly numbered: the first century C.E. comprises the years 1–100, the second 101–200, etc. So, the 20th century is generally agreed to be the years 1901–2000, making 2001 the first year of the 21st century. This did not,

The Chinese calendar is based on the lunar year, and works on a 60-year cycle. There are 12 months of alternately 29 and 30 days (with an intercalary month for leap years). Years are named after animals; the current cycle began in 2020 with the Year of the Rat, and ends with 2031's Year of the Pig.

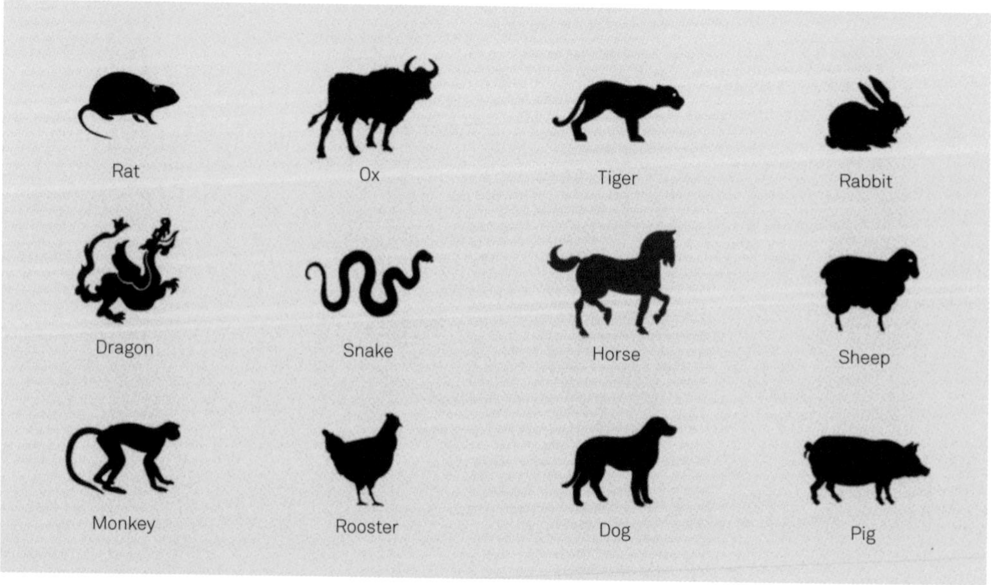

however, prevent worldwide celebration of the beginning of the second millennium at 00:00 on January 1, 2000.

millennium
a period of 1,000 years. We are currently in the third millennium C.E., but there is controversy over when this began (*see* **century**).

equinox
the time of year when day and night are of equal length; specifically the two occasions annually when the sun apparently crosses the celestial equator. They occur around March 21 (in the northern hemisphere the vernal equinox, and the southern hemisphere the autumnal equinox) and September 23 (in the northern hemisphere the autumnal equinox, and the southern hemisphere the vernal equinox). In astronomy the term is used to define the points where the sun crosses the celestial equator, the vernal equinox as it crosses south to north, the autumnal as it crosses north to south.

solstice
one of the two occasions each year when the sun is farthest north or south of the Equator. In the northern hemisphere, these correspond to the shortest day of the year, and longest day respectively; in the southern hemisphere vice versa.

quarter day
one of the four days which divide the year into quarters for mainly financial purposes, such as the payment of rent or interest. Traditionally in the U.K. the quarter days are Lady Day (March 25), Midsummer Day (June 24), Michaelmas (September 29), and Christmas Day (December 25).

Greenwich Mean Time (G.M.T.)
the standard time at longitude 0°. G.M.T. was adopted as the official standard time in the U.K. in 1880, and named after the Royal Greenwich Observatory in London. It subsequently became the basis of time zones around the world, but has been replaced by Universal Time (U.T.), also calculated from the time at 0° longitude. A further refinement to this system of time measurement is Co-ordinated Universal Time (U.T.C.), which uses international atomic time (T.A.I.) to fix time very precisely in SI **seconds**, making adjustments to compensate for irregularities in the Earth's rotation by adding leap seconds whenever necessary.

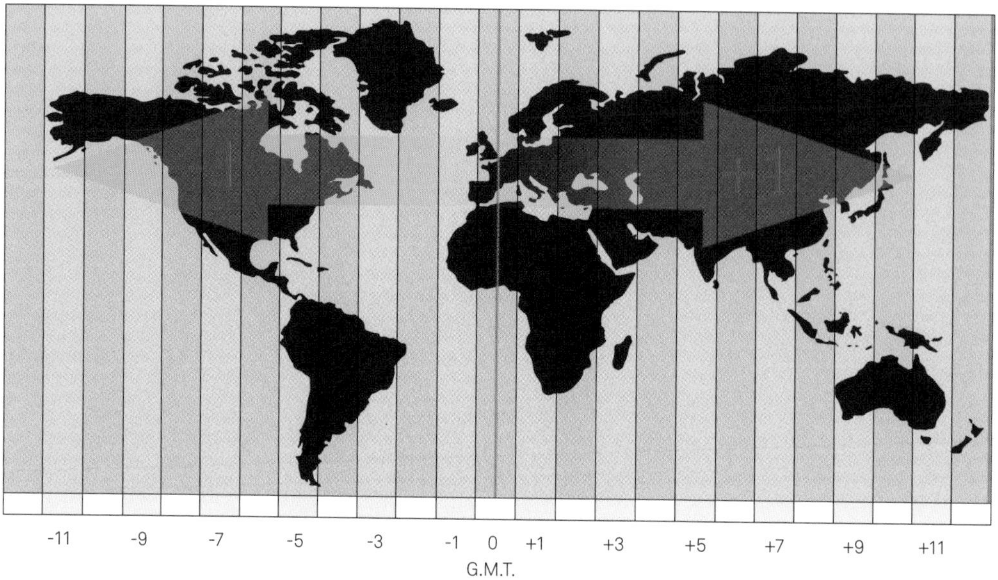

To minimize the number of time zones in any one country, the divisions do not follow lines of longitude exactly. Some countries, notably China, have a standard time despite spanning several time zones; others use fractions of an hour difference from U.T.

time zone

one of 24 divisions of the world having a standard time in all of its area. The divisions are essentially longitudinal segments of the globe of 15° each, 12 to the east of the prime meridian (longitude 0°), and 12 to the west, with some irregularities where countries cross the divisions. Time zones to the east are ahead of Universal Time, and are designated as positively-numbered, such as Central European Time which is time zone +01; the zones to the west are behind U.T., such as U.S. Pacific Standard Time which is time zone −08. The International Dateline follows, more or less, the line of longitude 180°, and marks where the positive and negative time zones meet. Crossing this line in an eastward direction takes you into the previous day, in a westward direction into the next day.

era

in calendrical usage, an era is a period of time in which a particular calendar has been in use. The starting date of any calendrical era is known as the emergent year. In terms of the **Gregorian calendar** in general use, we are now in the common era (C.E.), formerly referred to as Anno Domini (A.D.). Dates previous to the emergent year of 1 C.E. are described as before common era (B.C.E.), formerly before Christ (B.C.). Calendars in other cultures have different emergent years: for example, the Jewish calendar is dated from 3761 B.C.E., the supposed creation of the world, years being designated *anno mundi* (A.M.), or years of the world;

the Islamic calendar is reckoned from the Hejirah, Mohammed's flight from Mecca to Medina, in 622 C.E., designated in years *anno hegirae* (A.H.).

Julian calendar

the calendar established during the reign of the Roman emperor Julius Caesar, after whom it is named. The Julian calendar was based on a 365-day year, with a leap year of 366 days every fourth year, thus making the Julian year exactly 365.25 days. This was accurate enough for it to survive from 46 B.C.E. to 1582 C.E., when the inaccuracy of about 11 minutes each year had accumulated to an unacceptable level, and the more accurate **Gregorian calendar** was adopted.

Gregorian calendar

the calendar adopted in 1582 C.E. to replace the **Julian calendar**, and named after Pope Gregory XIII, who decreed its use. As the year is actually about 365.242 days, the 365.25-day year of the Julian calendar progressively got out of step with the seasons, and a new system was called for. Pope Gregory decreed that 10 days be dropped from the year 1582 to make up for this anomaly, and that to prevent future inaccuracy leap years should not occur in years divisible by 100, unless they are also divisible by 400, adding up to 146,097 days every 400 years, effectively defining the Gregorian year as exactly 365.2425 days—near enough for most practical purposes.

The Jewish calendar works on a 19-year cycle, using six lengths of year: common years of 353, 354 or 355 days, and leap years of 383, 384 or 385 days. The New Year (Rosh Hashanah) falls between September 5 and October 5 in the Gregorian calendar. The Islamic calendar is based on the lunar year, and works on a 30-year cycle; leap years occur every 2nd, 5th, 10th, 13th, 16th, 18th, 21st, 24th, 26th, and 29th years.

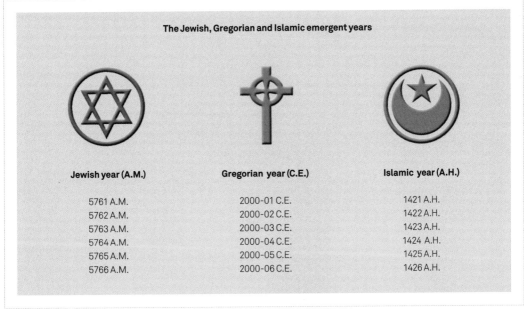

The Jewish, Gregorian and Islamic emergent years

Jewish year (A.M.)	Gregorian year (C.E.)	Islamic year (A.H.)
5761 A.M.	2000-01 C.E.	1421 A.H.
5762 A.M.	2000-02 C.E.	1422 A.H.
5763 A.M.	2000-03 C.E.	1423 A.H.
5764 A.M.	2000-04 C.E.	1424 A.H.
5765 A.M.	2000-05 C.E.	1425 A.H.
5766 A.M.	2000-06 C.E.	1426 A.H.

haab

the civil calendar used by the Mayans. The year was divided into 18 periods, called *uinal*, of 20 days, plus an extra five days called the *uayeb* to make a 365-day year (*see* facing page).

long count

the calendar used by the Mayans for defining historical dates, representing the number of days from the beginning of the Mayan era (possibly September 6, 3114 B.C.E.). The basic unit of the long count is the *kin* (day), and other components of the written date are multiples of either 20 or 18 days.

ancient Egyptian calendar

a calendar used by the ancient Egyptians, based on 12 months of 30 days plus an extra five days. This 365-day year, despite attempts to introduce a leap year, continued in use until around 25 B.C.E. Uniquely, the calendar was not based on the solar year or the lunar month, but instead on the nearly coincident rising of the star Sirius and the annual flooding of the river Nile, an important event for the agriculture of the region. Because of the calendar's gradual divergence from the tropical year, however, it soon became practically useless in marking the seasons.

The long count, used by the Mayans, is made up of five component parts. Longer periods, such as the *calabtun*, *kinchiltun* and *alautun* have also been used, but these are not included in the long count.

Components of the long count calendar

1 kin		1 day	
1 uinal	20 kin	20 days	
1 tun	18 uinal	360 days	c. 1 year
1 katun	20 tun	7,200 days	c. 20 years
1 baktun	20 katun	144,000 days	c. 394 years
1 pictun	20 baktun		c. 7,885 years

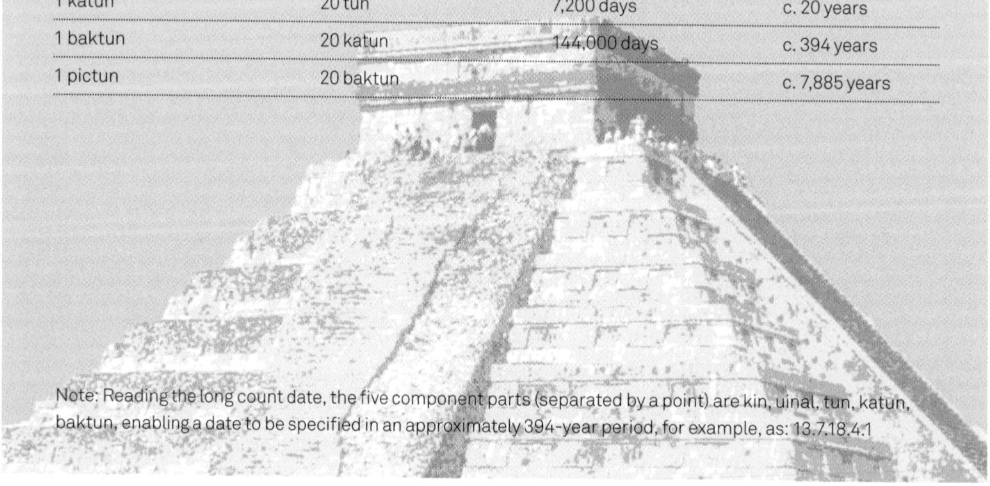

Note: Reading the long count date, the five component parts (separated by a point) are kin, uinal, tun, katun, baktun, enabling a date to be specified in an approximately 394-year period, for example, as: 13.7.18.4.1

Living things

livestock unit (LU)
the definition of livestock unit (LU) varies between different countries, and even between different areas within countries, but there is a rough general consensus that one large animal, e.g., a cow or a horse, amounts to 1 LU.

cow-calf unit
under the system of livestock units, a cow and calf are regarded as one animal until the calf is weaned, and are known together as a cow-calf unit. As such, they usually equal 1 LU, although under some definitions, they amount to 1.2 LU.

gestation period
the total time from conception to the birth of a viviparous mammal (meaning one giving birth to living offspring that have developed within the uterus). This period varies enormously according to species: the gestation period of humans, for example, is around 266 days; rats 21 days; cats 63 days and elephants 624 days.

Gestation periods for farm animals. Generally, the smaller the animal, the shorter the gestation period.

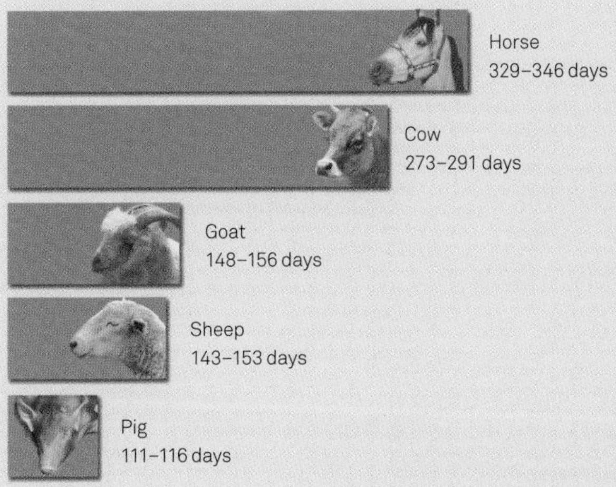

incubation period

the time that an egg takes to hatch, warmed either by the heat provided by the mother bird or in an incubator. The incubation period varies according to the time of year in which the eggs are laid. In medicine, the term also means the time between exposure to infection and the first signs or symptoms of a disease.

chromosome number

a chromosome is a threadlike body found in the nucleus of a cell. Chromosomes carry nucleoprotein in the form of genes that pass on hereditary characteristics. The chromosome number is the number of chromosomes in each somatic cell of an organism, which is constant for any one species. (The reproductive cells have half this number of chromosomes.) In humans, there are 23 pairs of chromosomes in a somatic cell, each pair containing one chromosome from each parent, so the chromosome number for humans is 46.

gene density

the number of genetic loci (positions) per unit of length on the linkage map. The genetic linkage map is the graphic representation of the order of genes within a chromosome, determined by how frequently two markers (or traits) are passed on between parent and child. Recombination of genes between chromosomes can alter the pattern of inherited characteristics.

allele

an allele is any one of a number of alternative forms of the same gene occupying a given locus (position) on a chromosome. These genes govern certain characteristics of an organism, but different alleles produce different forms of these characteristics. For example, different alleles produce different-colored petals in a flower.

centiMorgan

a centiMorgan (or map unit) is the distance between two genes for which the incidence of recombination is 1 percent. The American geneticist Thomas H. Morgan and members of his laboratory were the first to observe that genes are arranged on chromosomes in linear fashion, and that genes occurring on the same chromosome are inherited as a single unit for as long as the chromosome remains intact. Genes inherited in this way are said to be linked. Morgan and his colleagues discovered that there was often a breakdown of such linkage, and that recombination of genes occurred between chromosomes. The further apart the genes were, the more recombination was likely to occur.

biodiversity
the existence of a wide range of plant and animal life in a particular natural environment. The failure to preserve an area's biodiversity is a serious problem for natural balance, leading to the extinction of some species. Loss of habitat, including deforestation, pollution and the draining of wetlands, and competition from introduced species can upset biodiversity. Farming methods also affect whether an area's biodiversity is maintained, enhanced or reduced.

species classification
a species can be defined as a group of closely related organisms that are capable of interbreeding to produce fertile offspring. A method of classification of species, or taxonomy, was developed by Carl Linnaeus in the 18th century. Each species has two Latin names, the first being the genus, or group of species, and the second is the species itself. Each species has now been further categorized by family, order, class, phylum and kingdom.

I.U.C.N. Extinction risk categories
there are eight categories on the International Union for the Conservation of Nature and Natural Resources 'Red List' of endangered species: Extinct, Extinct in the Wild, Critically Endangered, Endangered, Vulnerable, Lower Risk, Data Deficient and Not

The system of species classification devised by Linnaeus has been extended. The table shows the categories for the blue whale.

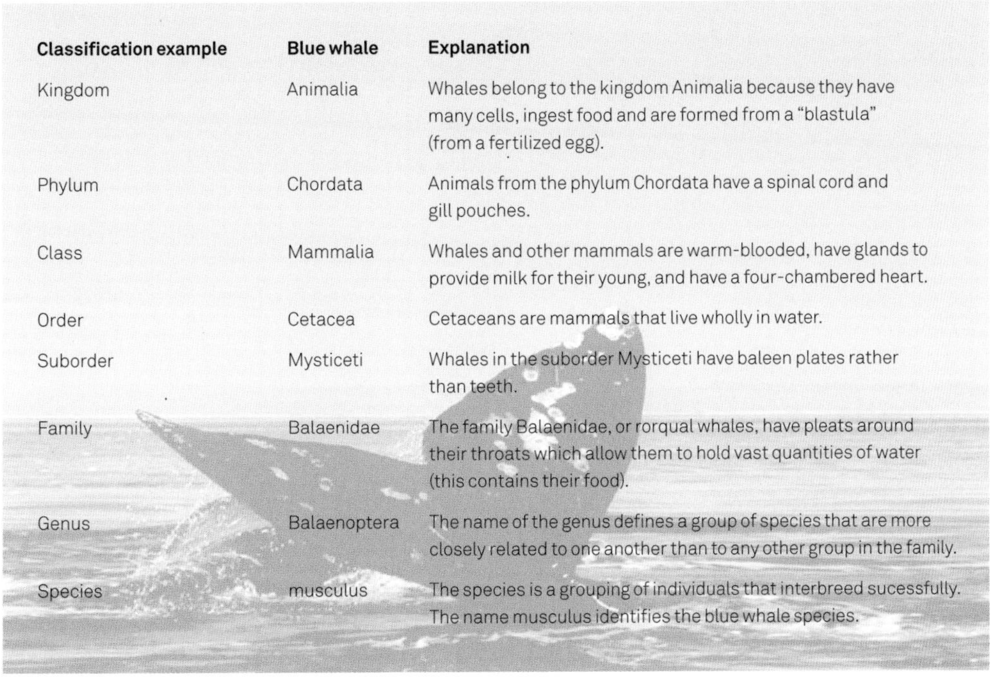

Classification example	Blue whale	Explanation
Kingdom	Animalia	Whales belong to the kingdom Animalia because they have many cells, ingest food and are formed from a "blastula" (from a fertilized egg).
Phylum	Chordata	Animals from the phylum Chordata have a spinal cord and gill pouches.
Class	Mammalia	Whales and other mammals are warm-blooded, have glands to provide milk for their young, and have a four-chambered heart.
Order	Cetacea	Cetaceans are mammals that live wholly in water.
Suborder	Mysticeti	Whales in the suborder Mysticeti have baleen plates rather than teeth.
Family	Balaenidae	The family Balaenidae, or rorqual whales, have pleats around their throats which allow them to hold vast quantities of water (this contains their food).
Genus	Balaenoptera	The name of the genus defines a group of species that are more closely related to one another than to any other group in the family.
Species	musculus	The species is a grouping of individuals that interbreed sucessfully. The name musculus identifies the blue whale species.

Evaluated. A species is described as "threatened" if it falls in the Critically Endangered, Endangered or Vulnerable categories.

population density

the number of people in a given land area, usually measured as the number per square mile or kilometer. It can be a rather misleading statistic: Canada has an overall population density of 3.4 people per square kilometer, but vast areas of the north have almost no people living in them, while the province of Ontario has a population density of 11.7.

population growth

population growth is usually measured annually by comparing the birth rate per 1,000 people and the death rate per 1,000 people.

clutch

the number of eggs laid by a single bird or laid in a single nest; or, sometimes, the brood successfully hatched from one laying of eggs. Hens lay clutches of five or more eggs; pigeons only two.

brace

a pair of game birds, as in the phrase "a brace of pheasants." Why game birds should be grouped together like this is not clear, though one plausible reason is that just one bird was considered inadequate for a meal, and two was about right. From a purely practical point of view, it is easier to string two birds together in order to carry them home.

Population density statistics can be misleading. Cities have high concentrations of people and inevitably have high population density compared to countries.

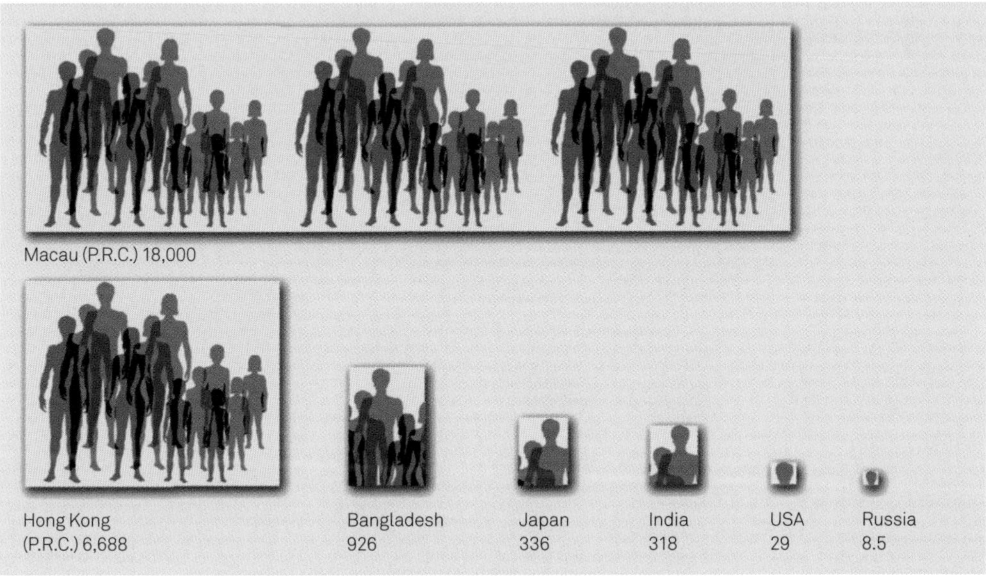

Physical sciences

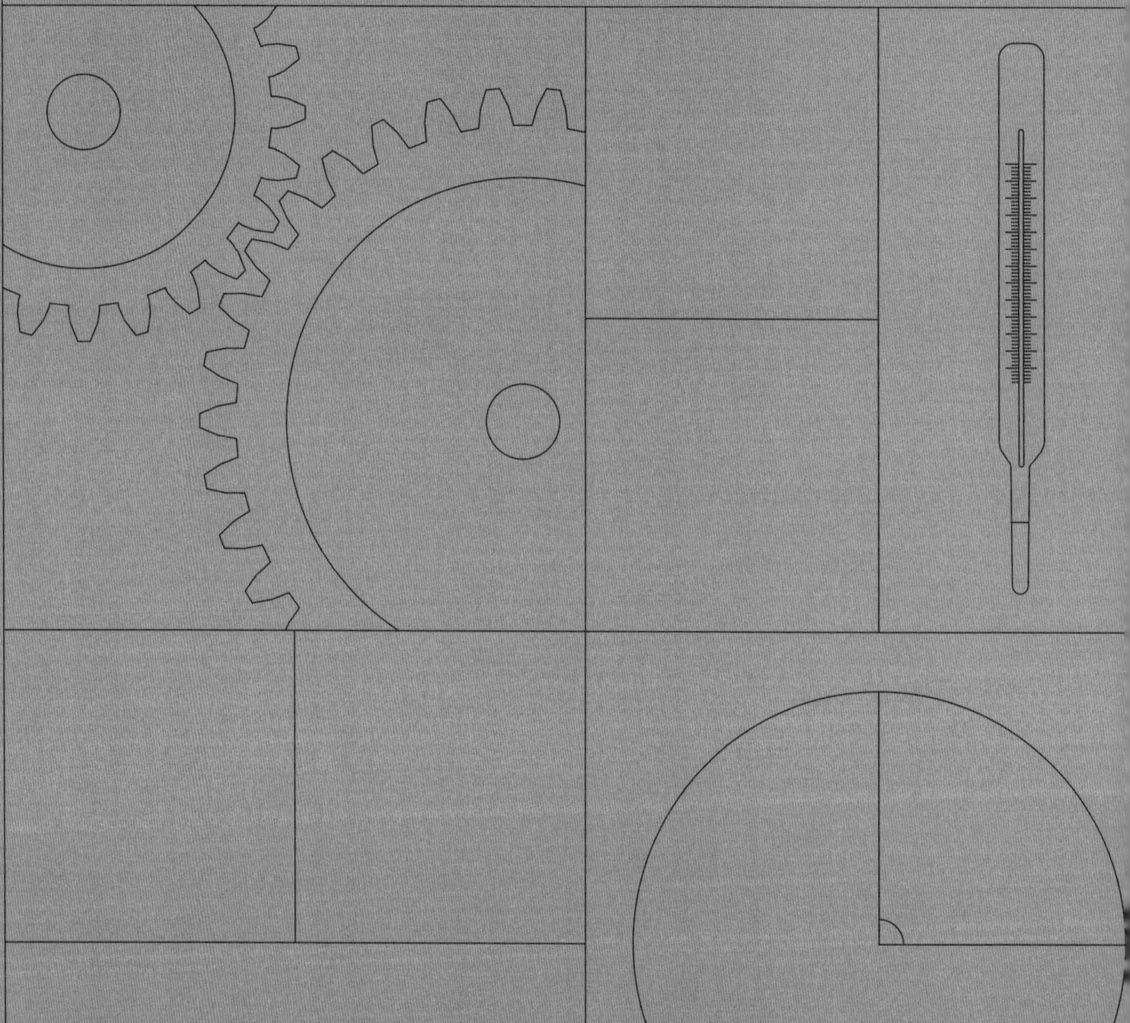

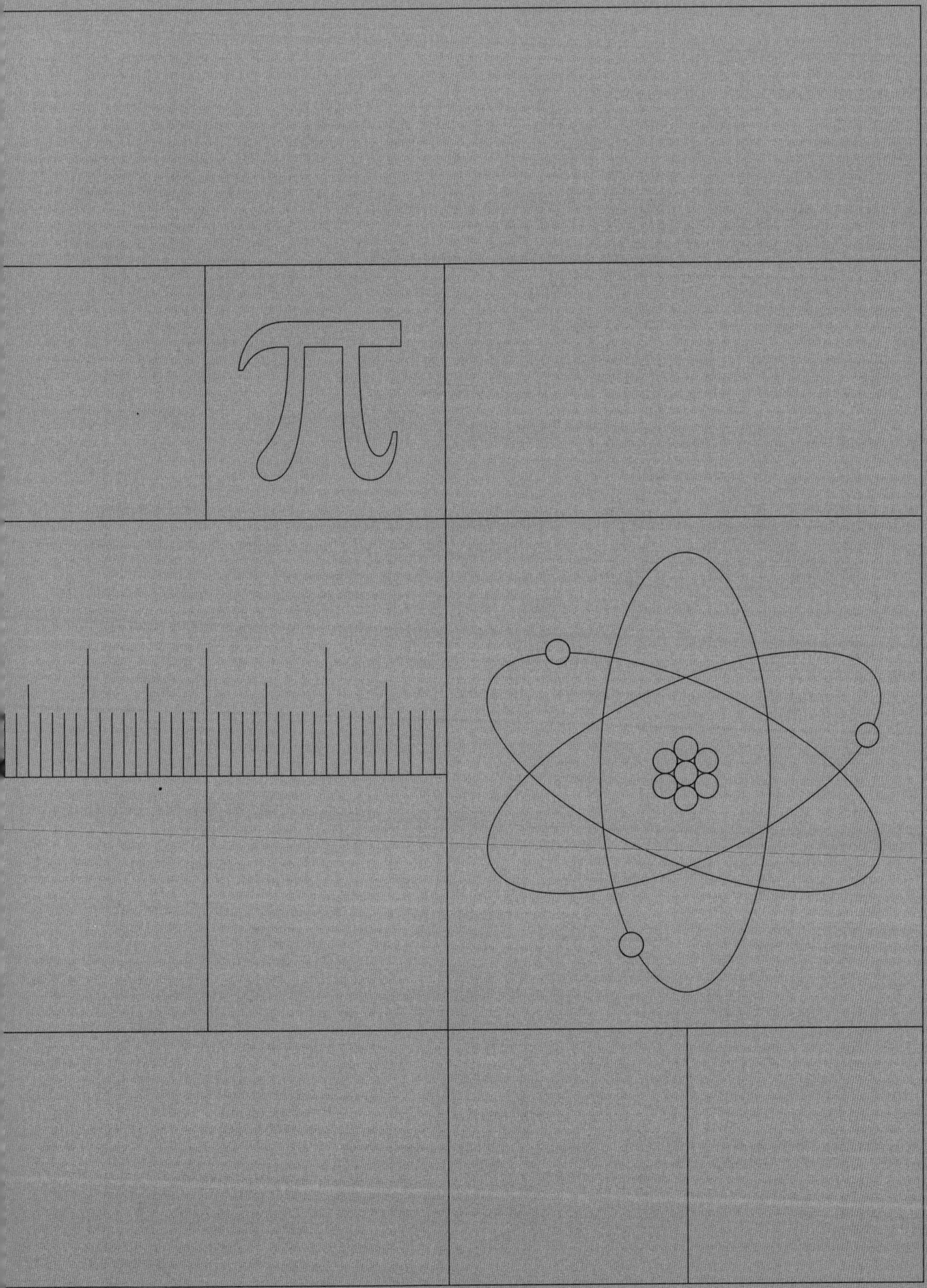

Chemistry

mole (mol)
an SI unit of quantity. One mole of a substance contains as many of its particles (typically atoms, molecules or ions) as there are atoms in 12 grams of carbon-12—approximately 6.022×10^{23} (a constant known as Avogadro's number). In practice, a mole is the number of atoms or molecules needed to make up the substance's atomic or molecular mass in grams.

relative molecular weight (or mass)
the weight of a molecule relative to $\frac{1}{12}$ the weight of a carbon-12 atom, calculated by adding the atomic masses (**atomic mass unit**) of each atom in the molecule. For example, water (H_2O) contains two hydrogen atoms (atomic mass 1.008 u) and one oxygen atom (atomic mass 16 u). So its relative molecular weight, is 18.016 u.

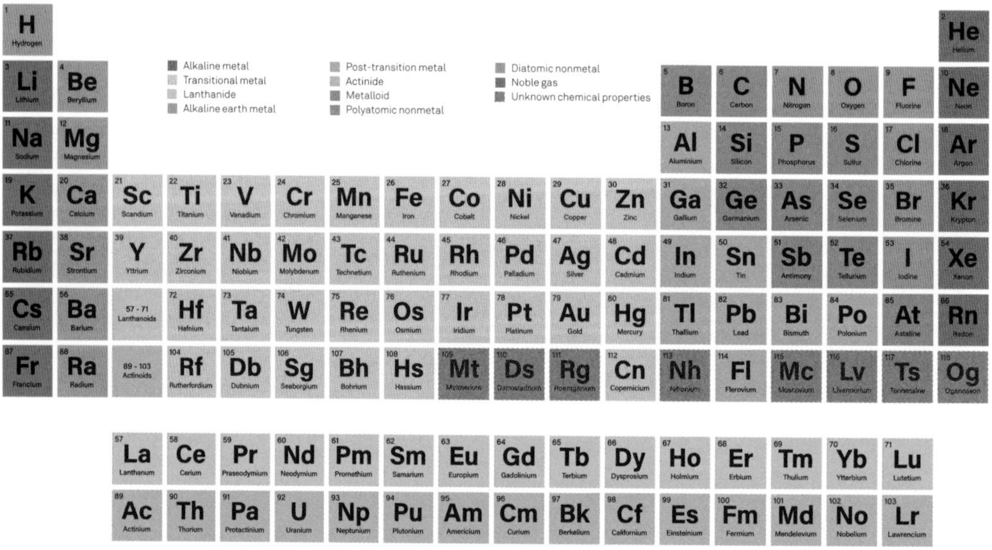

The periodic table has undergone a number of transformations ever since it was first suggested. This is the version currently in use.

periodic table 👁

a table of all known elements, arranged in increasing order of **atomic number**. For each one, the periodic table traditionally displays the atomic number, the symbol of the element and its relative atomic mass. The table began to take its current form in 1869 when Dimitri Mendeleev (1834–1907) arranged elements by atomic mass and predicted the existence of elements that were previously unknown.

period

a set of elements occupying a single row in the **periodic table**. As one moves from left to right, the atomic number—the number of protons in the nucleus—of the element increases by one. Elements in the same period also have the same number of electron shells, though the number of electrons in the electron shell of an element also increases by one as you move from left to right across the period. Elements horizontally adjacent to each other in the periodic table have a similar relative atomic mass, but they tend to have different properties.

group

a set of elements occupying a single column of the **periodic table**. There are 18 groups, or columns, in the periodic table, and elements that fall within the same group tend to have similar properties due to having the same number of electrons in their outer electron shell. For example, group 18 is populated by the noble gases helium, neon, argon, krypton, xenon and radon, which all show the similar property of being relatively unreactive.

isotope 👁

atoms of a particular element that have the same number of protons (which govern their place in the periodic table) but a different number of neutrons. Carbon-12, for example, has 6 protons and 6 neutrons, while carbon-14 has 6 protons and 8 neutrons. The variation causes isotopes of given elements to display different degrees of nuclear stability and makes some—known as radioactive isotopes—probe to nuclear decay.

allotrope

any of the different physical forms in which an element may occur. Allotropes of oxygen include O_2, a pair of bound atoms, and O_3 (ozone) where three oxygen atoms are bonded together. Allotropes of carbon include diamond, a three-dimensional crystal where each carbon atom binds to four neighbors, and graphite, a layered structure where each atom has strong bonds to three neighbors within the same layer.

The isotope of carbon-12 (top) has 6 protons and 6 neutrons. The isotope of carbon-14 (bottom) has the same number of protons, indicating that it is still carbon, but 8 neutrons.

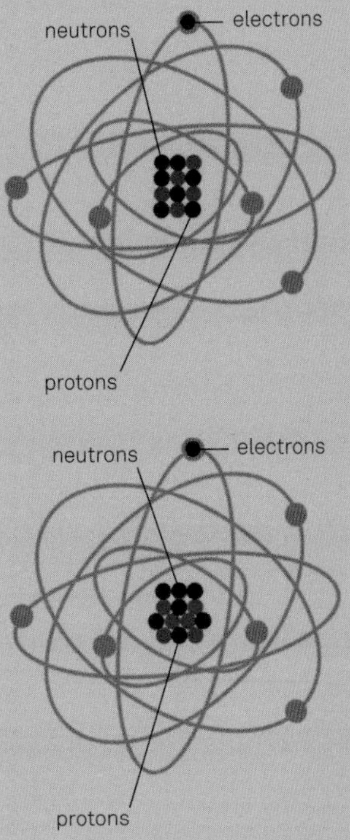

Litmus paper will indicate the pH of a particular solution by turning a certain color when it comes in contact with that solution.

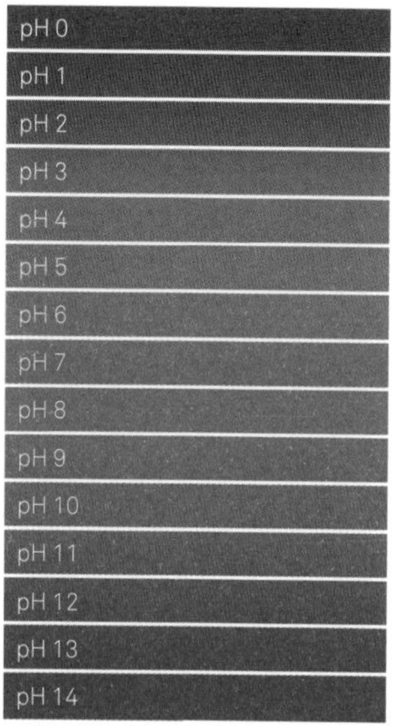

pH 0	battery acid
pH 1	sulfuric acid
pH 2	lemon juice, vinegar
pH 3	orange juice, soda
pH 4	acid rain (4.2–4.4) acidic lake (4.5)
pH 5	bananas (5.0–5.3) clean rain (5.6)
pH 6	healthy lake (6.5) milk (6.5–6.8)
pH 7	pure water
pH 8	sea water, eggs
pH 9	baking soda
pH 10	Milk of Magnesia
pH 11	ammonia
pH 12	soapy water
pH 13	bleach
pH 14	liquid drain cleaner

valency

a measure of the power of an element to combine with other elements to form a compound, specifically its power to combine with an atom of hydrogen. An element's valency is equal to the number of spaces left in the electron shells of its atom. So, the valency of oxygen is 2, because it requires 2 electrons to fill up its outer electron shell.

bond energy

the amount of energy released when a chemical bond is formed, equivalent to the amount of energy required to break that bond.

dipole moment

a measure of the extent to which electrical charge is separated with a molecule. Although the overall charge of a molecule may be neutral, this differential can allow magnets or electrical currents to exert a torque on the molecule.

electronegativity

a measure of the tendency of an atom to attract electrons and so form a covalent bond (where electrons are shared with other electronegative atoms) Elements toward the upper right of the periodic table tend to have higher degrees of electronegativity. The most electronegative element is fluorine; the least electronegative is francium.

pH

a measure of the acidity or alkalinity of a solution, according to its concentration of hydrogen ions. pH stands for potentiality of hydrogen, and is equal to $-\log_{10}C$ (**logarithm**) where C is the concentration of hydrogen in moles per liter. A pH of 7 indicates that a solution is neutral. A higher pH indicates a greater degree of alkalinity; a lower pH indicates a greater degree of acidity.

octane number

the percentage of octane in an octane–heptane mixture. Isooctane, for example, has an octane rating of 100. The octane number of a fuel is a measure of how smoothly that fuel burns.

katal (kat)

the SI unit of catalytic activity. A catalyst is a substance that increases the rate of a chemical reaction, without undergoing any chemical change itself. It is defined as having an activity of 1 katal if it allows a reaction to proceed at a rate of 1 mole per second.

$$2NO_2(g) \rightleftharpoons N_2O_4(g)$$

A chemical reaction describing the reversible reaction between nitrogen dioxide and dinitrogen tetroxide. The (g) stands for gas; many reactions are reversible in some states but not others.

reactivity
the rate at which an element or other substance tends to undergo a chemical reaction. The reactivity of a substance is essentially determined by its atomic or molecular structure, but it can be increased or decreased by changing its physical characteristics—e.g., grinding a substance to a powder may increase its reactivity.

reversibility
the ability for the products of a reaction to undergo a reverse reaction and recreate the original reactants. When a reversible reaction occurs in a confined space, an equilibrium is reached where the quantity of reactants and products does not change.

degrees of freedom
any of the ways in which an individual particle may move. Each degree of freedom of a particle will contain on average the same amount of thermal energy, equal to the temperature multiplied by the fundamental Boltzmann constant. Heisenberg's uncertainty principle states in quantum physics states that the amount of energy in any degree of freedom is never equal to zero.

buffer solution
a solution which resists change to its **pH** when a small amount of acid or alkali is added to it. Often consisting of a weak acid and its salt, buffer solutions maintain pH as a result of the acid and the salt reacting and entering a state of equilibrium. A solution of carbonic acid and bicarbonate exists as a buffer solution in blood plasma to maintain a pH of about 7.4.

molarity (M)

a measure of concentration. Molarity is the proportion of a substance in a solution, and is defined in terms of **moles** per liter.

molality (m)

a measure of concentration. Molality, like molarity, is the proportion of a substance in a solution, but defined in terms of moles per kilogram.

osmotic pressure

a measure of the force per unit of area required to stop the process of osmosis through a semi-permeable membrane. If the pressure exerted on the area of high concentration is increased relative to that exerted on the area of low concentration, the osmotic process will slow down and eventually stop.

rate of diffusion

the rate at which the natural and spontaneous spreading of particles occurs, e.g., smoke spreading in air or a drop of colored liquid in water. Osmosis is an important form of diffusion for humans as it is the process by which water enters the cells of the body.

Osmosis is a process whereby the solvent from a solution with a relatively high concentration of solute passes through a semi-permeable membrane to a solution with a relatively low concentration of solute. Such a membrane is permeable to the solvent, but not to the solute. The process continues until the solution is of the same concentration on both sides of the membrane.

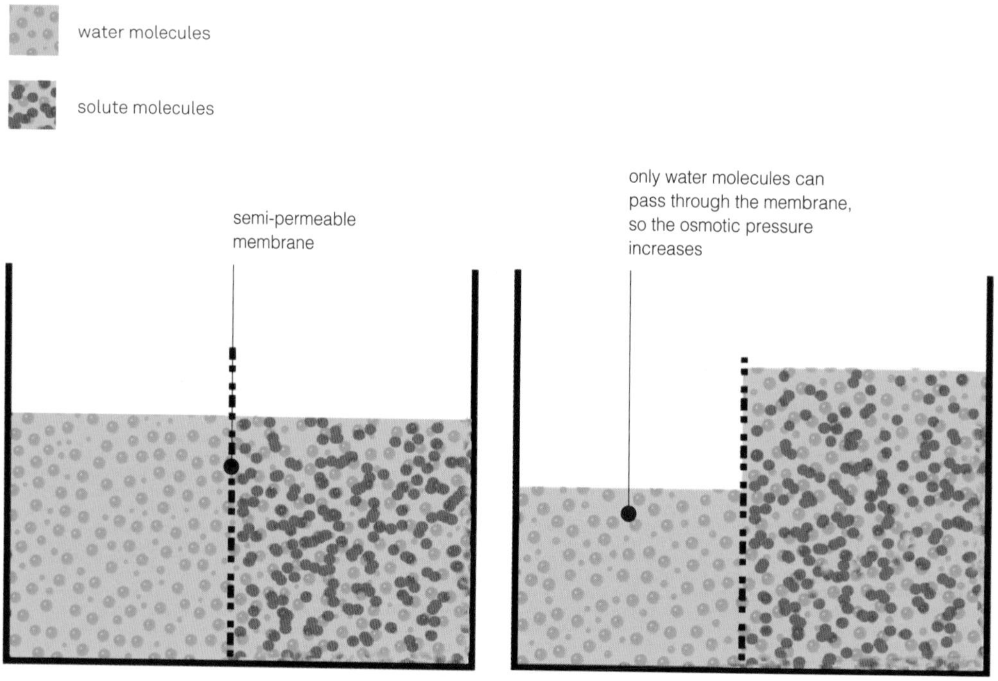

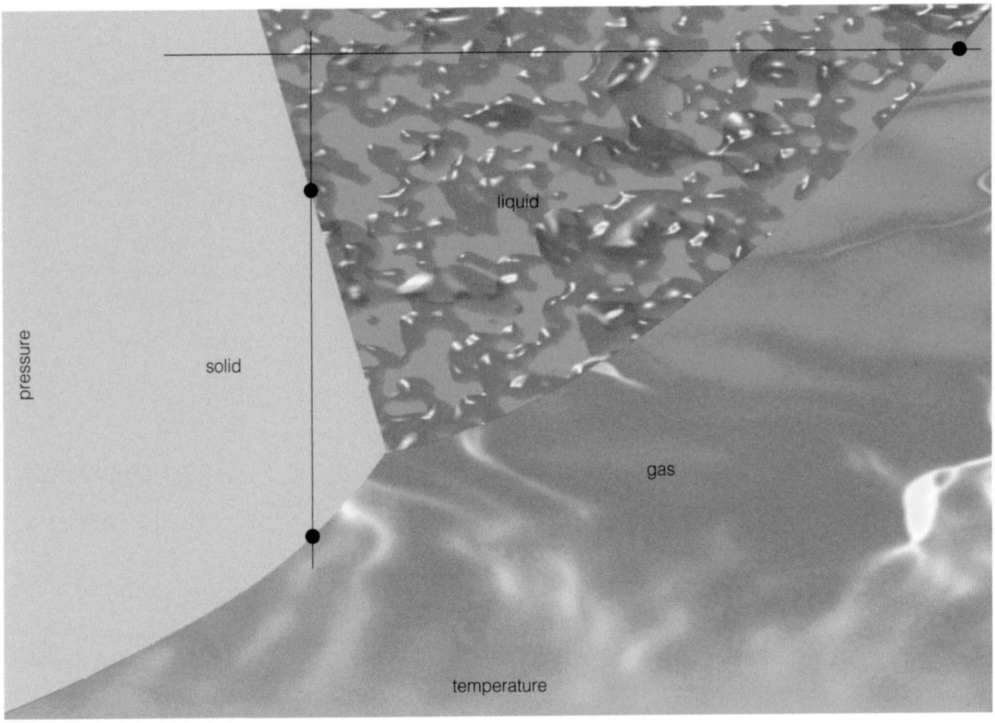

This phase diagram of water shows how temperature and pressure affect the state of matter of a substance. Using the phase diagram for a given substance, therefore, you can predict what state of matter the substance will be at if you know its temperature at a given pressure. Water is most unusual in that it melts under (immense) pressure.

diffusion gradient
any measure of diffusion in a given direction. If there is no gradient, there is no apparent change in concentration.

lattice types
any of the various formations which the atoms of a crystal may take. There are 14 different lattice types, which include the face-centered cubic lattice (fcc), the body-centered cubic lattice (bcc) and the sodium chloride lattice (NaCl).

unit cell
the smallest group of atoms in a crystal which display the overall symmetry of that crystal. The unit cell is repeated in three dimensions to form the lattice of a crystal.

phase diagram
a graph which displays the effect of both temperature and pressure on the three states of matter of a substance, or the effects of temperature and composition on a mixture of two (occasionally three) different substances. The first type is divided into three sections, one for each state of matter—solid, liquid and gas— or more (if the substance has **allotropes**).

Electricity and magnetism

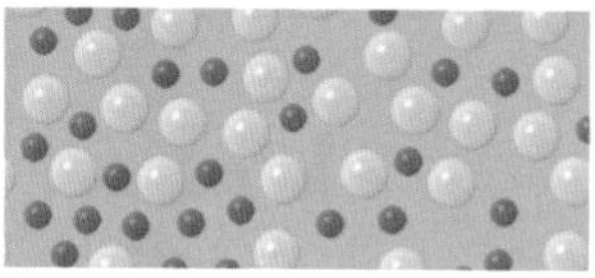

Resistance depends on the ability of electrons to move within a conductor. The greater the diameter of a wire, the greater the electron (current) flow.

resistance

the ability of a material to resist an electric **current** passing through it. When the material is a conductor, resistance is the ratio of the **potential difference** between the ends of the conductor to the current flowing through it. Resistance therefore equals the potential difference in **volts** divided by the current in **amperes**.

superconductor

the term for a material with a very high level of conductivity at a very low temperature. The specific resistance of metals decreases with temperature, and at a given "transition temperature," usually approaching absolute zero (–273°C), the electrical resistance of certain substances becomes zero. Hence, a current flowing through such a metal at this temperature will continue almost indefinitely.

ohm

the unit of electrical resistance, named after German physicist Georg Simon Ohm (1789–1854), who formulated Ohm's Law: the **potential difference** (in **volts**) across a conductor is equal to the **current** through it (in **amperes**) multiplied by the **resistance** (in ohms).

resistivity

also called "specific resistance," the **resistance** of a unit cube of a substance. This is usually a measurement in ohms per cubic meter (the ohm meter), and it is constant for a given substance at a given temperature.

current

the passage of electricity through a conductor. In a solid, usually a metal, the carriers of electricity are electrons. Electrons actually flow from negative to positive, but for historical reasons conventional circuit diagrams show them flowing from positive to negative. Current is measured in amperes. It is related to electrical power and **potential difference**: 1 watt is the energy expended per second by a steady current of 1 ampere flowing in a circuit across which there is a potential difference of 1 volt.

direct current (D.C.)
an electric current that flows in one direction only. Normally only used in low-voltage battery-operated devices.

ampere ("amp")
the SI base unit of **current**, usually shortened to "amp," named after French physicist André Marie Ampère (1775–1836). It was defined by the Conférence Générale des Poids et Mesures (which defines SI units) in 1948 as "that constant current which, if maintained in two straight parallel conductors of infinite length, of negligible circular cross-section, and placed one meter apart in a vacuum, would produce between these conductors a force equal to 2×10^{-7} newtons per meter of length." In 2019, the unit was redefined in terms of fixed fundamental constants of nature, but this has no practical effect on its value or usage.

amp hour (ampere hour; Ah)
the amount of energy charge in a battery that will allow 1 ampere of **current** to flow for 1 hour. Milliampere hours (mAh) are often used to describe the capacity of rechargeable batteries.

alternating current (A.C.)
a current that periodically reverses its direction, generally in a cyclical fashion. This means that the flow of electrons is constantly reversing. A.C. is the mains power supply of most countries, in most cases with a frequency of 60 cycles per second (60 hertz), but sometimes at 50 Hz.

root mean square (RMS)
the sum of the square roots of a series of values, divided by the number of values. In electricity, this has particular relevance to A.C. circuits: the RMS value, also known as the effective value, is the mean power level of an A.C. circuit. It is calculated from instantaneous values over a complete cycle. In the U.S. the RMS value is 110 volts, in Britain it is 240 volts.

hertz (Hz)
the SI unit of frequency, equal to one cycle per second. The frequency of a cycle is particularly relevant in the case of an

The lower of the two waves has a shorter wavelength but a higher frequency measured in hertz, than the wave at the top.

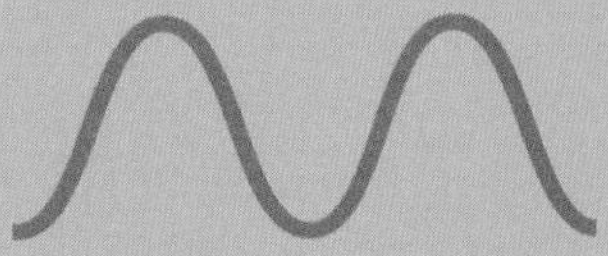

long wavelength, low frequency, low energy

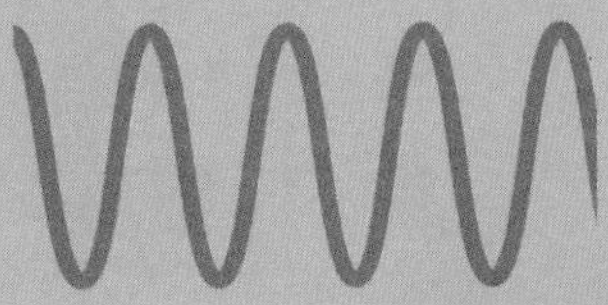

short wavelength, high frequency, high energy

Electrolysis is used for electroplating objects, some types of purification and performing certain chemical reactions. The anode accepts electrons (either from atoms attached to it, which then become positive ions, or from negative ions in the solution, which become neutral atoms) while the cathode neutralizes positive ions in the solution by donating electrons to them. In electroplating, it will be metal ions that form a very thin coat of solid metal on the surface of the cathode.

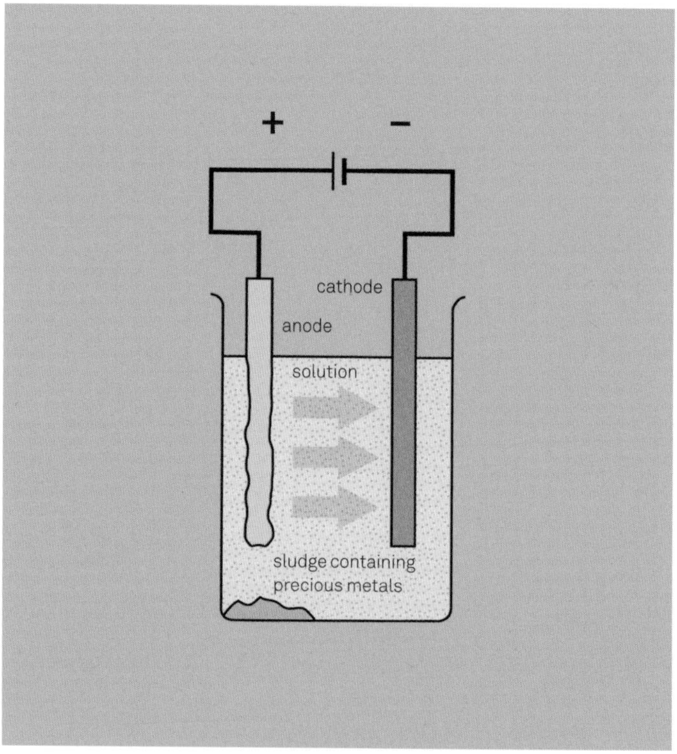

alternating current 👁, and the electromagnetic waves produced by it. Electromagnetic radiation is the basis of wireless communication, and includes radio and infrared waves. Since the speed at which electromagnetic waves move is fixed (equivalent to the speed of light), their wavelength is inversely proportionate to their frequency in Hertz. So the higher the wavelength, the lower the frequency. The unit is named after German physicist Heinrich Hertz (1857–94).

potential difference
the voltage in a circuit. If work is done to pass a charge between two points, then a potential difference exists between them, and this is measured in **volts**.

volt
the unit of **potential difference** or electromotive force. One volt is the potential difference between two points in a circuit when 1 **joule** of energy is expended in making 1 **coulomb** of electricity pass from one point to the other. One volt can also be expressed as using 1 **watt** of power at a constant current of 1 **ampere**, or as the potential difference across a resistance of 1 ohm when a

current of 1 ampere is passed through it. The unit is named after Italian physicist Alessandro Volta (1745–1827).

electron-volt (eV)
the increase in energy of a single electron falling through a **potential difference** of 1 **volt**, used as a meausre of energy in atomic and nuclear physics. One electron-volt = 160.206×10^{-21} joules, and other convenient units are the mega-electron volt, giga-electron volt and even the tera-electron volt. The Large Hadron Collider, the world's largest particle accelerator, has achieved collision energies of up to 13 TeV.

Faraday's constant
the amount of electric **charge** required to deposit 1 **mole** of any mono-valent element, equal to about 96,485 coulombs. The unit is also known simply as Faraday (symbol F), and is named after British electrochemist and physicist Michael Faraday (1791–1867), who first identified it. Faraday's constant is particularly important in electrolysis. His first law of electrolysis states: "The amount of a substance deposited on each electrode of an electrolytic cell is directly proportional to the current through the cell."

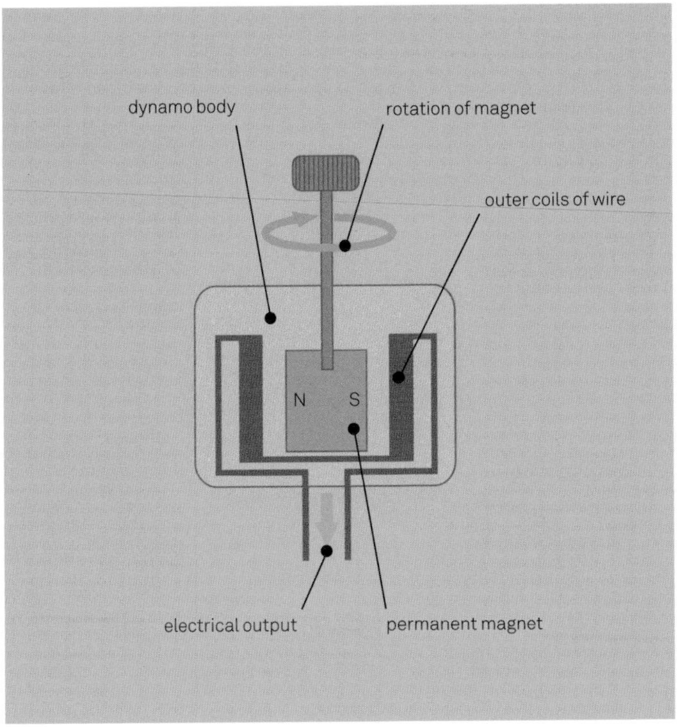

The dynamo uses magnetism to provide electromotive force, and hence current.

Magnetic moment is described mathematically as m = IA, where I is the current and A is the area enclosed by the current loop.

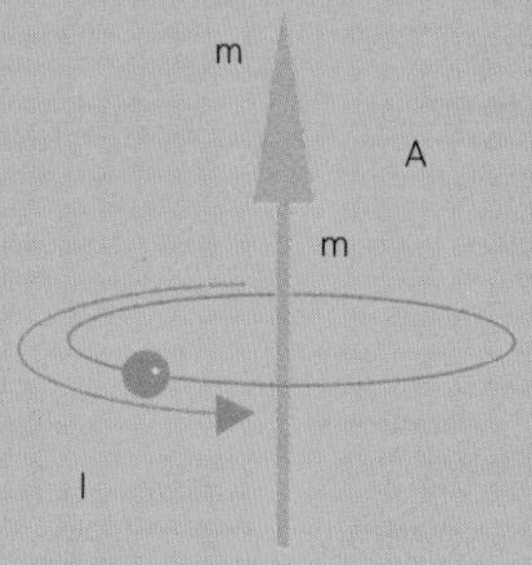

charge

the existence of an excess or deficiency of electrons in a body. A body with an excess of electrons is said to be negatively charged, and one with a deficit positively charged. Charge may be quantified in terms of coulombs.

coulomb

the SI unit of electrical or electrostatic charge: the amount of electricity transferred when a current of 1 **ampere** flows for 1 second. Named after French physicist Charles Coulomb (1736–1806), who formulated Coulomb's Law. This law states that the mutual force between two point electrostatic charges is proportional to the product of the charges, and inversely proportional to the square of the distance between them. One coulomb = 6.3×10^{18} elementary charges (an elementary charge is the charge on one electron or proton).

capacitance

the ability to store an electrical charge. A capacitor is an electronic component usually with two plates separated by a small distance. When a voltage is passed across the capacitor, a positive charge is stored on one plate, and an equal negative charge on the other. The capacitance, measured in farads, is the ratio between the charge (measured in coulombs) and the voltage.

inductance

the property of an electric circuit opposing any change in electric current. A change in a circuit causes an electromotive force (E.M.F.) opposing the change, measured in henrys (E.M.F. in volts, per ampere per second current change).

eddy current

induced current created in the core of an electromagnet or transformer by changing magnetic fields. The heat produced is an indication of inefficiency, or wasted energy, and measures can be taken to reduce it. One way is to laminate the metal core, with insulators placed between each thin sheet of metal.

gauss

a unit of magnetic induction or magnetic flux density (magnetic field) in the C.G.S. system. One gauss is 1 **maxwell** per square centimeter. A magnetic field of 1 oersted produces an induction of 1 gauss in air, and of μ gauss in a medium whose magnetic permeability is μ. One gauss is equivalent to 10^{-4} **tesla**.

magnetic flux density

the magnetic flux per unit area, measured in **webers** (joules per ampere). Magnetic flux density is measured in **teslas** (webers per square meter).

maxwell

a unit of magnetic flux in the C.G.S. system. It is the flux per square centimeter, perpendicular to a magnetic field with intensity of 1 **gauss**. The unit was named after British physicist James Clerk Maxwell (1831–79).

tesla

the SI unit of **magnetic flux density**, amounting to one weber per square meter. Named after Serbian-American physicist Nikola Tesla (1846–1943).

magnetic field strength (H)

also known as "magnetic intensity" and "magnetizing force," and measured in **amperes** per meter. It is defined as a **vector** quantity whose magnitude is the strength of a magnetic field at a point in the direction of the magnetic field at that point. It is proportional to the length of a conductor and the amount of electrical **current** passing through the conductor.

magnetic moment

for a magnetic field of unit magnetic field strength, the torque that is required to hold a magnet perpendicular to the field. It is also expressed as the product of the pole strength of a magnet and the magnetic length (the distance between the poles). Electrons and thus atoms have magnetic moments, some of which are very small, but others, like those of iron, are very large. The size of an electron's magnetic moment is expressed in Bohr magnetons, named after Danish physicist, Niels Bohr (1885–1962).

Temperature

absolute zero

the temperature at which an object will have no heat energy at all, and its constituent atoms therefore reach what is called their "ground state" (they can never actually stop moving completely). Absolute zero is (in theory) the lowest possible temperature. In practice, the laws of thermodynamics make it impossible ever to reach absolute zero, but scientists have managed to come within a billionth of a **kelvin**.

Celsius (°C) 👁

a temperature scale, with 0°C fixed at the freezing point of pure water, and 100°C at its boiling point (both at standard atmospheric pressure). It was originally called centigrade, but was officially renamed in 1948 to honor Swedish astronomer Anders Celsius, who invented it in 1742.

The Fahrenheit, Celsius, kelvin and Rankine temperature scales.

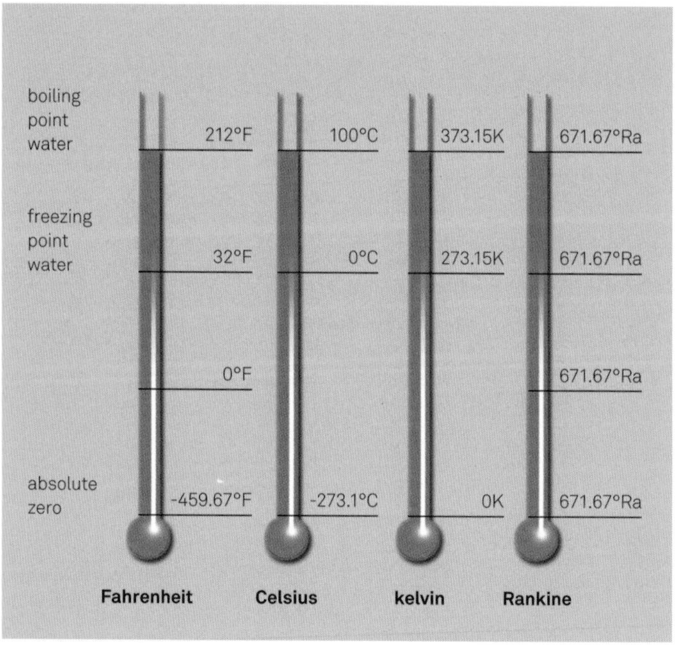

kelvin (K)]

the SI scale of temperature. One kelvin is equivalent to 1° Celsius, but 0 K is set at absolute zero, equal to –273.16°C (so water freezes at 273.15 K). The full name of the unit (named after Britishscientist William Thomson, later Lord Kelvin) should be written with a lower-case k, but the abbreviation is a capital K and neither is used with "degrees" or the ° symbol. In 2019, the kelvin was redefined in terms of fundamental constants of nature, but this has no practical effect on its value or usage.

Fahrenheit (°F)

a temperature scale proposed in 1724 by German physicist Gabriel Fahrenheit. His initial scale was calibrated from the melting point of a mixture of equal weights of salt and water at 0°F, and the temperature of horse blood (assumed to be the same as a human's) at 96°F, and the melting and boiling points of water at 32°F and 212°F, respectively. The scale was recalibrated after his death using the freezing and boiling points of pure water as the fixed points.

Rankine (°Ra)

a rarely used scale with the same degree size as Fahrenheit but (like the kelvin scale) with 0°Ra at **absolute zero**. Named after the Scottish physicist William Rankine, who proposed it in 1859.

infrared

the part of the electromagnetic spectrum between microwaves and visible light (between about 300 GHz and 400 THz). It is emitted by any object with a temperature above **absolute zero**, and is felt by a human as heat.

boiling point

the temperature at which a substance changes from liquid to gas (also called the condensation point, since the reverse transition from gas to liquid occurs at the same temperature). It varies considerably with pressure.

freezing point

the temperature at which a substance changes from liquid to solid (or vice versa, so that it is sometimes called the melting point). Most substances freeze more easily under high pressure. Water, however, behaves rather unusually; it expands when cooled below 4°C, so that ice melts under pressure.

Night temperature	Tog
15°C to 8°C	3–5
10°C to 0°C	5–8
3°C to −10°C	7–10

Note: Sleeping bags used by backpackers outdoors have their tog values equated to a comfort temperature rating.

sublimation point
a few substances (such as the element iodine) pass directly from solid to gas on heating (and vice versa on cooling), and cannot normally exist in a liquid state. This process is called sublimation, and the temperature at which it occurs is the sublimation point.

triple point
the combination of temperature and pressure at which some substances can exist simultaneously in gaseous, liquid and solid forms, all in equilibrium. (The triple point of water is at 0.01°C and 612 Pa, with no other gas present.)

Standard Temperature and Pressure (S.T.P.)
a standard to allow replication of scientific experiments in controlled, identical conditions. It is taken to be 0°C and 101,325 Pa. A similar concept, R.T.P. (Room Temperature and Pressure), may be used to mean a temperature of about 20°C and ambient atmospheric pressure—far easier to achieve without special equipment.

specific heat capacity (c)
the amount of energy needed to raise the temperature of 1 kg of a substance by 1 K, without changing its state.

latent heat
The amount of energy needed to change the state of 1 kg of a substance, through a specific transition, without altering its temperature. Energy must be added to change from liquid to gas or solid to liquid, while the opposite changes require energy to be removed.

temperature gradient
the rate at which the temperature changes with distance across a material (including air). Easily calculated for a single substance by dividing the total temperature difference by the distance from one side to the other.

thermal conductivity
a measure of a substance's ability to allow heat to pass through. Calculated by multiplying the flow rate of heat energy by the thickness of material, and then dividing this product by the cross-sectional area of material multiplied by the temperature difference.

thermal diffusivity
the ratio of a substance's thermal conductivity to its specific heat capacity. Thermal diffusivity is a measure of how quickly an object will match the temperature of its surroundings.

frigorie
a unit of refrigeration, defined as the extraction of heat at a rate of 1 kilogram calorie per hour. The name is modeled on "calorie," using the Latin *frigus* (cold) instead of *calor* (heat), but the unit is too small for most practical purposes.

tog 👁
a unit of thermal resistance, defined as the temperature difference in degrees Celsius (between the warm and cold sides) divided by 10 times the flow of heat energy (in watts per square meter) from one side to the other. Used mainly for bedding and winter clothing, typical tog values range from about 5 (for lightweight summer bedding) up to 15 (a very heavy winter duvet).

Clo
another unit of thermal resistance, also used primarily for textiles, 1 clo is a little over 1.5 tog. One clo is the amount of insulation that maintains a heat-flow of 50 kilogram calories per hour between 70°F on one side and body temperature on the other.

Total heat energy graph for converting ice at −100°C to steam at 200°C. About two-thirds of the total energy is used in the phase change converting water at 100°C to steam at 100°C.

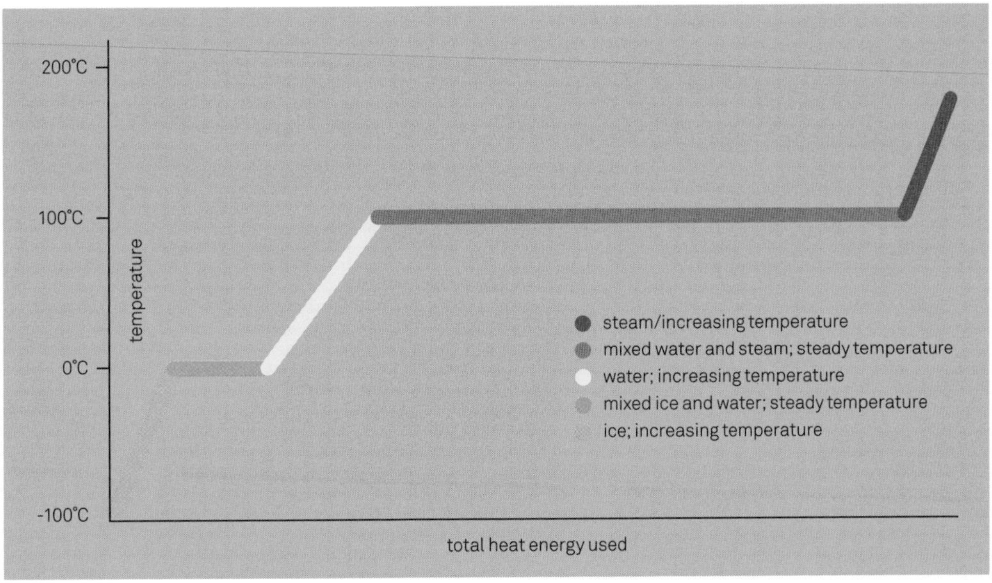

U-factor (U-value)
a unit of thermal conductivity for building materials, measured as the heat loss in **Btu** per square foot per hour for a temperature gradient of 1°F per unit thickness. The lower the U-value, the better the material works as an insulator. The equivalent unit of thermal resistivity is the R-value, equal to 1 divided by the U-value.

R.S.I. value
the SI unit of thermal insulation, measured in kelvin meters-squared per watt ($K.m^2.W^{-1}$).

Btu (British thermal unit)
technically a unit of energy, but used primarily in the context of heat (including central heating systems and steam turbines). Defined as the amount of energy required to heat 1 lb of water by 1°F, although the starting temperature of the water being heated changes the value of the Btu slightly (the higher the temperature, the lower the energy in 1 Btu). One Btu is around 1,055 joules, so one Btu per hour is a little under 0.3 watts. A therm is equal to 100,000 Btu.

thermometer
a device for measuring temperature. The most common type is a sealed glass tube with regular markings, containing a liquid which expands steadily on heating (usually mercury or dyed alcohol). Other types include thermistors (electronic devices that vary their resistance with temperature) and **thermocouples**. A thermograph is a thermometer that keeps a constant record of temperature—for example, by drawing a graph or feeding data to a computer.

thermostat
a feedback system used to control temperature (for example, in fridges or boilers). It consists of a thermometer (often a thermocouple) and an automatic control system (either electronic or mechanical) for the source (or extractor, in a fridge) of heat.

thermocouple
a cheap, simple and reliable form of thermometer, a thermocouple consists of two lengths of different metals connected at one end, with an instrument for measuring voltage joined to the other ends. Changing the temperature at the joint causes a proportional change to the voltage across the metals.

pyrometer
a thermometer designed for use at very high temperatures, e.g., in pottery kilns, industrial processes such as the blast furnace, or within a volcano.

entropy
a measure of the amount of energy (in a collection of objects) that is unavailable to do work. In thermodynamics, entropy is defined as the difference between the amount of energy in the system (collection of objects) and the amount of energy available to do work divided by the absolute temperature. It is not necessarily the same as heat energy, since some heat energy can be transformed into other forms of energy (and thus made to do work), provided there is a temperature gradient. It is never possible to convert all of a system's heat energy to work, however (which is why perpetual motion machines are impossible). The laws of physics suggest that total entropy can only increase—your fridge can reduce the entropy of its contents, but only by increasing the entropy of its surroundings by a greater amount.

Entropy can also be a measure of the extent to which a collection of objects (atoms in a gas or liquid, stars in a galaxy, files on a hard drive, etc.) is disordered. This is statistical entropy, where the likelihood of a specific state occurring by chance is considered—if you flip 100 coins, you might wind up with 100 heads, but this is extremely unlikely as there is only one possible arrangement (low entropy). You are far more likely to get 99 heads and one tail (100 possible arrangements), and the most likely (highest entropy) result is 50 of each (large numbers of possible arrangements: 10^{29} of them, in fact).

Light

R.G.B. (red, green, blue)
the human eye contains three different types of cone-shaped light receptors, which detect three different ranges of wavelengths. Other animals see other ranges of wavelength, and can have different numbers of "cones." Different colors trigger different combinations of cones (e.g., red triggers only the yellow-sensitive cones, yellow triggers a mixure of yellow- and green-sensitive cones), allowing a full spectrum of colors to be detected. A similar system of mixing three colors of light is used for computer and T.V. screens. An intensity is assigned to each of red, green and blue, which combine to give the desired color. Most current computer graphics systems use 8 bits for the intensity of each of the three colors, (24 bits in total) allowing for 16.7 million colors.

C.M.Y.K. (cyan, magenta, yellow, key (black))
a color model used by printers. The system is subtractive: the more of each color added, the less light is reflected. Small dots of ink are printed in each color (variations in dot size affet the intensity of that color), and from a distance blur to give the appearance of a single color. Black is added as a separate color because, while in theory combining all the primaries at maximum intensity should give black, the practical result tends to be murky brown.

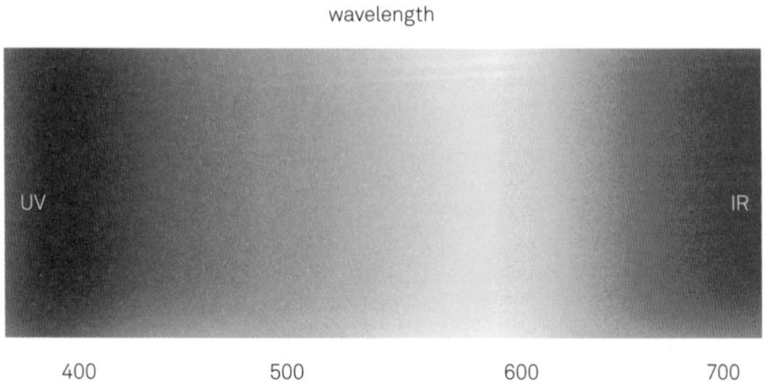

Humans can detect the range of light spectrum from about 400nm to about 700nm. We see this as a smoothly varying rainbow of colors, from violet (higher frequency) to red (lower frequency).

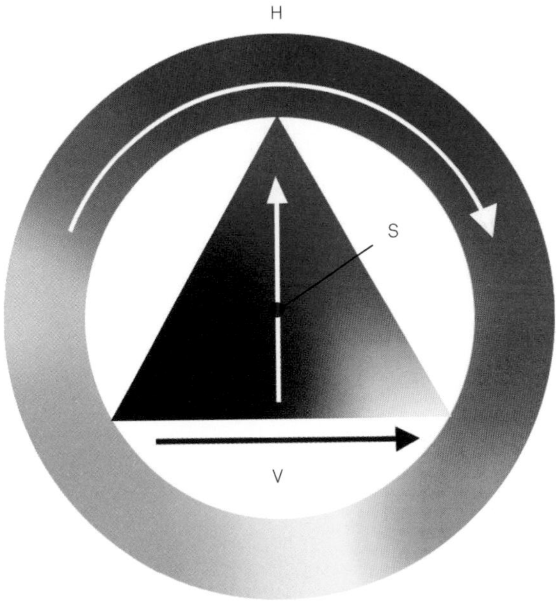

A standard representation of the H.S.V. color space.

Hue = red, green or blue.
Saturation = from achromatic white to pure hue.
Value = brightness, from black to pure hue.

H.S.V. (hue, saturation, value)
a color model used mainly by artists to mix opaque paints, though it can be used to describe colors in other media, including some computer graphics programs.

hue
the element of a color that most people would think of as "color." It is usually described as a continuous wheel, with the spectrum along the rim, and red shading into purple.

saturation (or chroma)
the vibrancy, or intensity, of a color. A low-intensity color will be a shade of gray with a hint of the **hue** in it. Pastels are very high-**value**, low-saturation colors. Measured as a percentage, 100 percent is a fully saturated color, and 0 percent is a pure gray. (Differs from **luminous intensity**.)

value (V)
a measure of how light or dark a color is (also sometimes called its brightness). Like **chroma**, measured as a percentage: pure black is V0 percent, regardless of chroma; pure white is V100 percent, regardless of **hue**. Some hues are perceived as intrinsically brighter than others and do reflect more light even at maximum saturation and value—yellow is the brightest color, and blue tones tend to appear the darkest.

luminous intensity
a measure of the energy emitted as light in a particular direction. A related but different measure is the radiant intensity, which is the amount of energy emitted as electromagnetic radiation (including light) in a particular direction. Because the eye responds differently to different colors of light, the same radiant intensity produces different luminous intensities for different wavelengths of visible light.

candela (cd)
the SI unit of luminous intensity. It is defined as the luminous intensity (in a given direction) of an object emitting (in the direction specified) $1/683$ watts per steradian of monochromatic electromagnetic radiation, with a frequency of 540 terahertz. The frequency is that at which the human eye is most sensitive (a slightly yellowish green), and the slightly random-seeming $1/683$ is to maintain the value produced by a previous definition involving a "black body" held at the melting point of platinum (around 1,770°C).

lumen (lm)
the SI unit of luminous flux, best understood as the total amount of light emitted into a solid angle of 1 **steradian** from a light source with an even intensity of 1 **candela**.

lambert
a unit of luminance, the **luminous intensity** of a surface, i.e., the amount of light that the surface emits or reflects per unit area. A surface with luminance 1 lambert emits (or reflects) 1 lumen per square centimeter.

lux (lx)
the SI unit of illumination of a surface, i.e., the light falling onto the surface. It is defined as 1 lumen per square meter. The illumination provided by most artificial light sources is extremely low, relative to natural sunlight—an office brightly lit with strip lamps will have an illumination of between 300 and 500 lux, while daylight ranges from around 30,000 lux (on a cloudy day) to over 100,000 lux.

diopter
a measure of the focusing power of a lens, equal to 1 divided by the focal length of the lens. For a converging lens, this is a positive number; for a diverging one, the number is negative.

polarization
the angle at which a **photon**'s electrical wave component is oriented. While it cannot be measured for an individual photon, the polarization of a light source can be established by measuring the angle at which a polarizing filter allows the maximum amount of light through. Most normal light sources are unpolarized (their individual photons have random polarization), but light can become polarized by passing through a filter or reflecting off a surface. Polaroid (and similar) sunglasses work on this principle, by absorbing light polarized horizontally, to reduce glare.

reflectivity
the ratio of energy reflected to that absorbed or passed through for light hitting a surface.

refractive index (n) 👁
a measure of the amount by which electromagnetic radiation (including light) is slowed down when traveling through a material relative to traveling in a vacuum. Calculated by dividing c (the speed of light in a vacuum) by the speed of light in the material in question. The refractive index determines the amount by which light passing into a material at an angle will bend.

diffraction
the ability of a wave, including light, to bend as it passes around the edge of a barrier. The extent to which a given wave will diffract depends on the relationship between the width of the gap through which the wave passed and its wavelength, peaking when the two are approximately equal.

As you move farther from a light source, the intensity of the light you receive from it will decrease.

1 meter 4,000 lux   2 meters 1,000 lux   3 meters 444 lux

optical density

a measure of an object's ability to absorb light. Each unit of optical density represents an order of magnitude decrease in the light passing through. That is, an object with an optical density of one allows 10 percent of light entering to pass through (absorbing the other 90 percent), while an optical density of two allows only 1 percent of light entering to pass through (absorbing 99 percent), etc.

laser

an acronym for Light Amplification by Stimulated Emission of Radiation. A device designed to produce highly focused, usually monochromatic, light, often in very short bursts. Since the beam of a laser is highly collimated (the light is all traveling in almost exactly the same direction), it cannot normally be seen unless an object is placed in the path of the beam to reflect it, which makes high-power laboratory lasers, able to cut through sheet metal (or a human being), extremely dangerous. The power of lasers in everyday use (primarily pointers and C.D. players) are typically measured in milliwatts. However, even these will cause permanent damage if pointed into an eye. Lasers are also useful for making a variety of measurements, because their light can be very tightly controlled.

The refractive index of a substance controls how widely separated are the red and violet fringes produced by a prism of that material. (The same effect produces rainbows in the sky.)

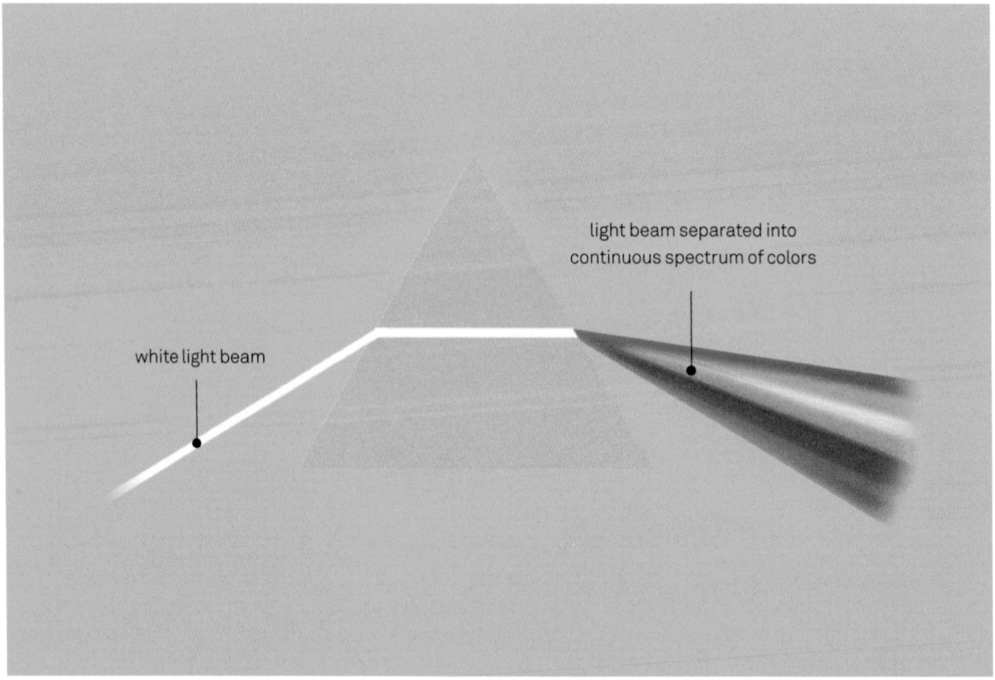

wavelength
the distance between two adjacent peaks of a wave (including light, sound and water waves), equal to the speed at which it travels divided by its frequency. For light, this is typically in the range of 400 nanometers (violet) to 740 nanometers (red).

photon
the quantum (smallest possible quantity) of light or any other electromagnetic radiation. A photon can be thought of as a self-reinforcing burst of electromagnetic and magnetic waves aligned at right-angles to each other and the direction of their motion through space. The energy of a photon in **joules** is equal to its frequency (in **hertz**) multiplied by the Planck constant (6.6×10^{-34}), and the laws of physics dictate that it remains in constant motion at the speed of light until its energy is absorbed and converted to another form—at which point the photon, which has no rest mass, ceases to exist.

rayleigh (R)
a very small unit of light intensity, primarily used for astronomy. Equal to one million photons per square centimeter per second.

standard erythemal dose (S.E.D.)
a measure of cumulative absorbed energy from ultraviolet light, defined as 100 joules per square meter (normally of skin surface). The related measure of intensity is expressed in S.E.D.s per hour, equal to 27.8 milliwatts of skin-affecting ultraviolet light per square meter.

Mathematics

gradient
a measurement of rate of change. Colloquially, the gradient is a measure of the steepness of a hill, and is defined in terms of a ratio. A 1:2 gradient increases or decreases one unit in height for every two units in length. In mathematical terms, a gradient is a **vector** acting on a scalar function at a given point in a scalar field. It may therefore exist in more than two dimensions.

vertex
a mathematical term having a number of similar meanings. Geometrically, a vertex is any of the angular points of a polygon or polyhedron. In this sense it sometimes used interchangeably with **apex**, but strictly speaking the apex is only one of the vertexes. A vertex is also the point at which an **axis** meets a curve.

axis
an imaginary straight line. In geometry, an axis is the imaginary straight line around which a plane is rotated to create a symmetrical solid. In a wider mathematical sense, an axis is a fixed line against which coordinates are measured, commonly named an x-axis, y-axis or z-axis. An axis of symmetry is the line in respect of which a two- or three-dimensional shape may be symmetrical.

The point of inflexion for this curve is that point where the gradient of the curve changes from being positive to negative.

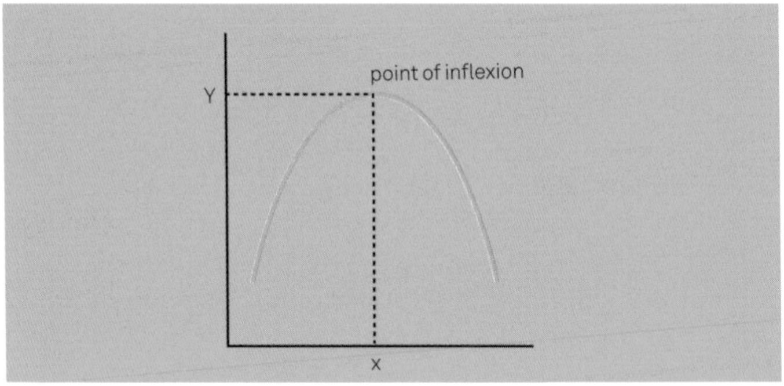

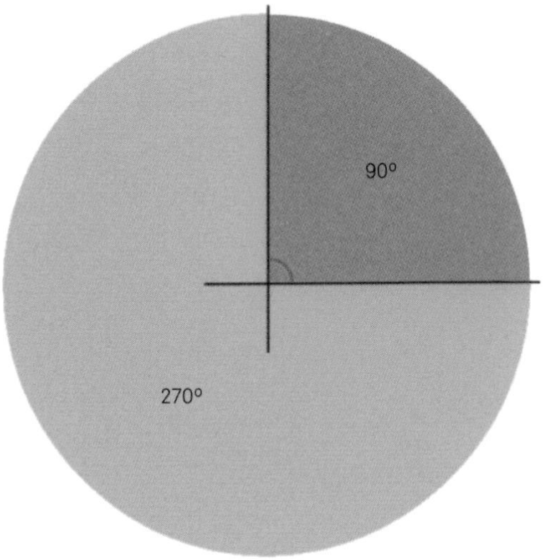

A circle is divided into 360 arc degrees, 32,400 arc minutes and 1,944,000 arc seconds.

point of inflexion (inflection)

the point at which a curve changes from being concave to convex. More specifically, a point of inflexion is the point at which the sign (positive or negative) of the rate of change of the gradient itself changes.

asymptote

a straight line that becomes continually closer to a particular curve, without ever meeting that curve.

arc degree (°)

a unit of angular measure, also simply known as a degree. There are 360° in a full circle, 180° in a semicircle, and 90° in a right angle. Arc degrees are ways of measuring the size of angles, but can also be used to measure lengths in relation to distance. For example, your thumb, held at arm's length from you, has a length of about 2°. Arc degrees are subdivided into arc minutes (‘) and arc seconds (“).

radian (rad)

the SI approved unit of angular measure, and an alternative to the **arc degree**. One radian is the angle subtended at the center of a circle by an arc of the same length as the radius of that circle. As such, there are 2π radians in a circle, and 1 radian is equivalent to approximately 57.3°. Radians are used in calculus to make results as natural as possible.

Steradians, being measures of three-dimensional angles, are always conical in shape. There are 4π steradians in a sphere.

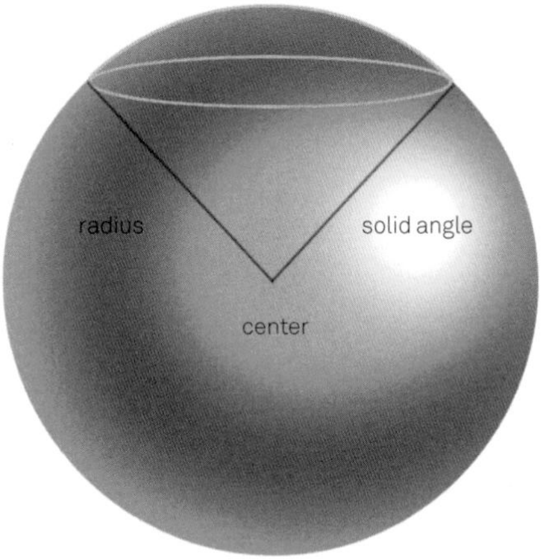

steradian (sr)

a unit of three-dimensional angular measure. The steradian is the SI unit of solid angle, and the three-dimensional equivalent of the **radian**. One steradian is the angle subtended at the center of a sphere by a part of the sphere whose surface area is equivalent to the square of the radius of the sphere.

sine

for one of the non-right-angle corners of a right-angle triangle, the sine is the length of the side opposite the angle as a fraction of the hypotenuse. The equivalent for the side adjacent to the angle is the cosine.

tangent (tan)

a mathematical term which has two distinct meanings. In geometry, a tangent is a straight line which touches a curve at a point where the curve and the straight line have the same gradient. The geometrical tangent of a curve is defined using calculus. In trigonometry, the tangent is the ratio of the sides other than the hypotenuse opposite and adjacent to an angle in a right-angled triangle.

integer

a whole number. The set of integers is that set of numbers which includes all the positive natural numbers (1, 2, 3, 4 ...), all the negative natural numbers (–1,–2, –3, –4 ...) and 0. Any number containing a fraction or a decimal is not an integer. The set of integers is indicated by mathematicians by the letter **Z**, which stands for *Zahlen*, the German for numbers. Number theory is the branch of mathematics that includes the study of integers.

real

a word used to describe any number that contains no **imaginary** part; indeed the term real was coined by mathematicians in response to the concept of imaginary numbers. In practical terms, a real number is any number that corresponds to a point on the infinite number line. Real numbers may be positive, negative, rational, irrational, algebraic or zero.

imaginary

the word used for the non-real part of a complex number that can be written in the form a + ib, where a and b are real numbers and **i** is the square root of −1. A complex number is one that has both real and imaginary parts—"ib" is the imaginary number. Imaginary numbers are central to many sciences and branches of advanced math.

prime

any natural number whose only positive divisors are itself and 1. In number theory, prime numbers are the basic building blocks of all natural numbers, which is to say that any integer can be expressed as the product of prime numbers. There are an infinite number of prime numbers, but the largest known prime is $2^{24,036,583}-1$, which has 7,235,733 digits. The opposite of a prime number is a composite number.

Definition of the tangent function. The unit circle is the circle with its center at the origin and a radius of 1. Angle x is formed by rotating OA about the origin to OP. Point Q is the intersection of line OP and x = 1. Then the y-coordinate of point Q is tan x.

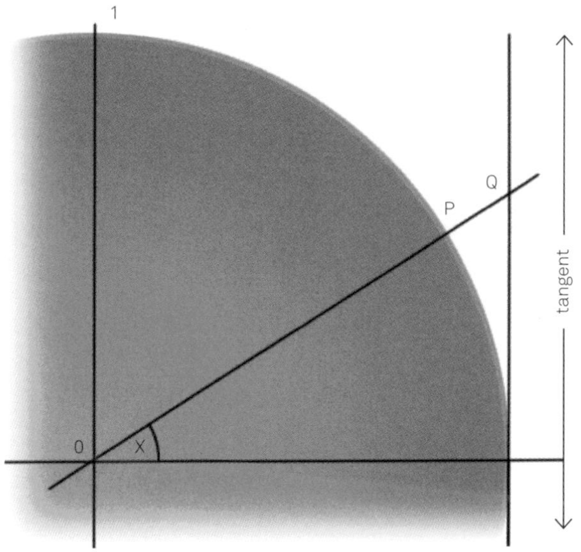

rational

a word used to describe any number that can be expressed in terms of the ratio of two integers. In other words, a rational number is any number that can be expressed in the form a/b, where a and b are integers and b does not equal 0.

irrational

a word used to describe any real number that cannot be expressed in terms of the ratio of two integers. Examples of irrational numbers are the square root of 2, **π** and **e**.

infinity (&)

a number greater than any assignable value. The concept of infinity has evolved over history, and has philosophical and cosmological as well as mathematical connotations.

i

an imaginary number equal to the square root of −1. The mathematical symbol i stands for imaginary unit, and it is properly defined as the solution to the equation $x^2 = -1$.

e

a mathematical constant equal to approximately 2.71828. The mathematical symbol e stands for Euler's number, and is named after the Swiss Leonhard Euler (1707–83). It is a transcendental number, and the base of the natural **logarithm** function.

Pi is perhaps the most important constant in mathematics. It is an irrational number, which means it has an infinite number of decimal places.

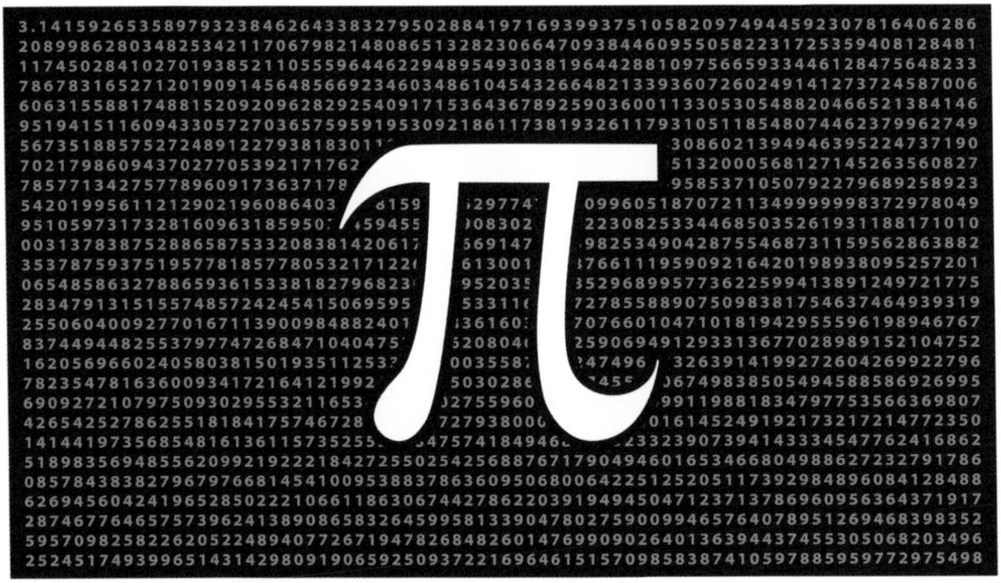

pi (π) 👁
a mathematical constant approximately equal to 3.1415927. Pi is defined as the value of the ratio of the circumference of a circle to its diameter, and is an irrational number. It is one of the most important mathematical constants, essential to describing circles: the area of a circle is πr^2, where r is the radius; and the circumference of a circle is $2\pi r$.

random
a statistical term used to describe events that happen in such a way that cannot be accurately predicted. This is not to say, however, that random events are unpredictable when taken en masse. For example, snowflakes may fall randomly, but it is possible to predict the general area in which they will fall, and the cumulative nature of their falling.

arithmetic sequence
a finite or infinite sequence of numbers in which each number is equal to the previous number plus a constant, also known as the "common difference." So, an arithmetic sequence starting with 3 and with the common difference 4 would begin 3, 7, 11, 15, 19 ...

geometric sequence
a finite or infinite sequence of numbers in which each number is equal to the previous number multiplied by a constant, also known as the "common ratio." So, a geometric sequence starting with 3 and with the common ratio 4 would begin 3, 12, 48, 192, 768 ...

exponential sequence
a finite or infinite sequence of numbers in which each number is equal to the previous number raised to the power of a constant exponent. So, an exponential sequence starting with 3 and with the exponent 4 would begin 3, 81, 43046721 ...

Fibonacci sequence 👁
an infinite sequence of numbers in which each number is equal to the sum of the previous two numbers. The simplest Fibonacci sequence is that which starts with the numbers 0 and 1, and begins 0, 1, 1, 2, 3, 5, 8, 13 ...

average—mean
an average figure derived from taking the sum of several quantities and dividing it by the number of quantities. Thus, the mean average of 1, 3, 5 and 7 is (1+3+5+7)/4 = 4. The arithmetic

The start of the Fibonacci sequence that begins 0, 1. Italian mathematician Leonardo Fibonacci (c. 1170–1250) is also credited with popularizing the use of Arabic numerals in Europe.

0 1 1 2 3 5 8 13
21 34 55 89 144
233 377 610 987

mean is often used, but it can be very misleading, as it can be skewed by unrepresentatively large or small figures at either end of the population.

average—mode
an average figure equal to the most common quantity in a number of quantities. For example, the modal average of 1, 2, 2, 3, 3, 3, 4, 5 is 3, as this is the figure that appears the most often.

average—median
an average figure below which 50 percent of the quantities in a population fall. If there is an odd number of quantities in the population, the median will be a number in the population. If there is an even number of quantities in the population, the median will be the mean average of the two central quantities in the population. So, the median average of 1, 2, 2, 3, 5, 6, 8, 8, 9, 10 is 5.5, the figure below and above which 50 percent of the figures fall.

average—midrange
the mean average of the highest and lowest quantities in a number of quantities. For example, the midrange average of 1, 3, 5, 6, 6, 6, 7, 11 is (11+1)/2 = 6.

probability
a means of estimating the likelihood of the occurrence of an event. In mathematics, probability is generally represented in terms of a real number between 0 (impossible) and 1 (certain). More colloquially, probability is represented in terms of fractions: the probability of a dice roll being 3, 5 or 6 is ½ (0.5). Gamblers represent probability in terms of odds, or ratios: a 2:1 chance of an event occurring is the equivalent of a ⅔ probability.

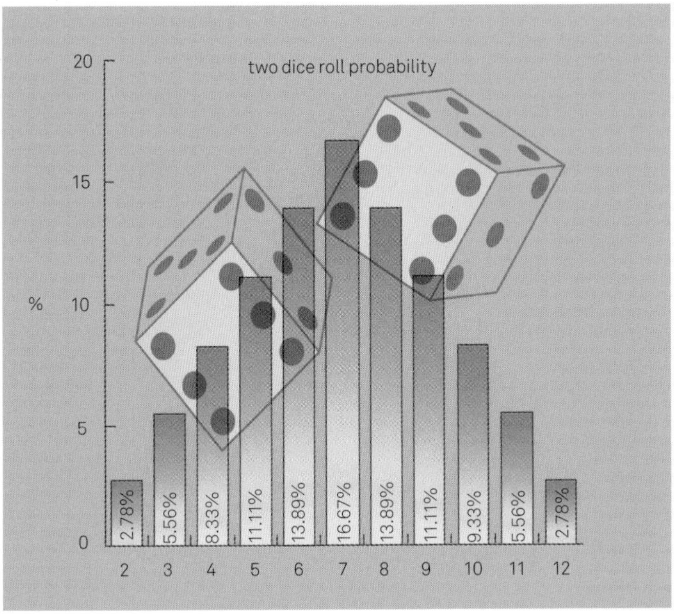

The probability of rolling a three is 1/6, but that does not mean that if you roll a dice six times you will definitely roll a 3. The probability of a given event is not affected by previous events. This graph illustrates the probability of getting a particular number when you roll two dice.

combination
a selection of a number of elements from a larger number of elements, without regard to the way in which those are ordered. For example, three combinations of three numbers may be derived from the sequence {1, 2, 3, 4}, namely {1, 2, 3}, {1, 2, 4} and {2, 3, 4}. However, there are six permutations of each of these combinations – distinct ways in which they can be ordered.

googol
a number equivalent to 10^{100}, or 1 followed by 100 zeros. The term was "invented" by a nine-year-old in 1938, Milton Sirotta, the nephew of American mathematician Edward Kasner (1878–1955).

googolplex
a number equivalent to 1 followed by a **googol** of zeros, or ten raised to the power of a googol. If the digits of a googolplex were written in 1-**point** type, its length would be 4.7×10^{69} times the diameter of the known universe.

standard deviation (σ; s)
a measure of statistical dispersion. The standard deviation of a set of numbers is the mean difference between those numbers and their mean. The mean of the numbers 1, 3, 4, 6, 7 is 4.2, and the standard deviation is approximately 2.387, indicating that on average the difference between each number and 4.2 is 2.387.

A large standard deviation tells us that on average the elements of a population are a long way from the mean. For example, the populations 29, 30, 31 and 20, 30, 40 both have a mean of 30, but the standard deviation of the latter is much larger than that of the former.

normalize

the process of multiplying a series, function or number by a factor such that some associated quantity—often, but not necessarily, the norm—is equal to a desired value, usually 1.

quartiles 👁

four equal groups into which a population of values of a particular variable can be divided. Each quartile represents a quarter of the population. The first quartile cuts off the lowest quarter of the data, and is also known as the 25th percentile. The second quartile cuts the set of data in half, and is also known as the median or the 50th percentile. The third quartile cuts off the highest quarter of the data, and is also known as the 75th percentile. The difference between the first and third quartile is known as the interquartile range.

percentiles

a hundred equal groups into which a population of values of a particular variable can be divided. If a figure appears in the 95th percentile, then it is in the top 5 percent of that population.

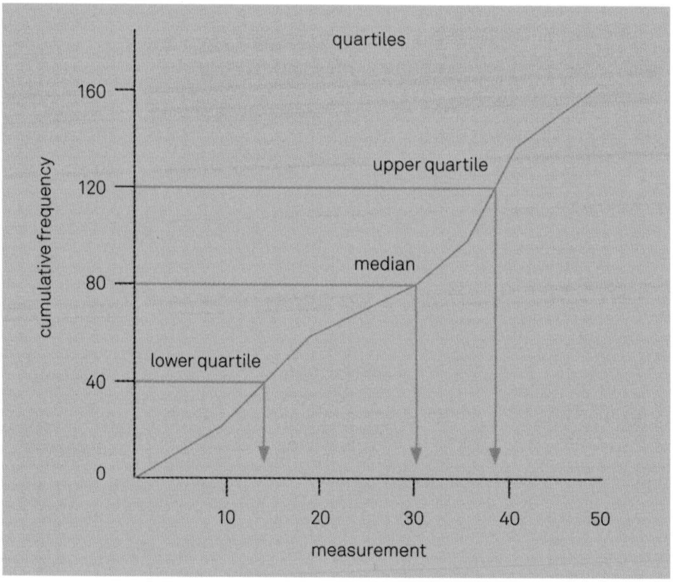

Quartiles are a type of percentile: the 25th or lower percentile, the 50th (median) percentile and the 75th or upper percentile.

percent (%)

a means of representing a decimal, fraction or proportion as a whole number. Percent derives from the pseudo-Latin phrase *per centum*, meaning out of a hundred, hence 25 percent represents the fraction $^{25}/_{100}$. Percentages can be greater than 100 percent—150 percent, for example, represents an increase of 50 percent. Put another way, 150 percent = $^{150}/_{100}$ or 1.5.

numerator

the number above the line in a vulgar fraction. The numerator indicates how many of the parts indicated by the **denominator** the value of the fraction equals. So, for example, in the fraction $^{2}/_{3}$, the numerator is 2, and it indicates that the fraction is equal to 2 one-third parts.

denominator

the number below the line in a vulgar fraction. The denominator indicates the number of parts into which the population the fraction describes is divided. So, in the fraction $^{3}/_{4}$, the denominator is 4, and it tells us that the population is divided into four equal parts, or quarters. Denominators are never equal to 0.

factors

integers by which another number is exactly divisible. For example, 1, 2, 3 and 6 are all factors of 6, because 6 can be divided precisely by any of them.

factor

a number by which another number is increased or multiplied. So, 10 increased by a factor of 4 is 40.

factorial (!)

the product of a positive integer and all the integers below it except 0. So, 6! is equal to $6 \times 5 \times 4 \times 3 \times 2 \times 1 = 720$. 0! is taken to be equal to 1. For any group of n objects, there are n! permutations (ways of ordering them), while the number of **combinations** of choosing k objects from among them is equal to $n!/k!(n - k)!$. Factorials are also used widely in calculus and probability theory.

bases

a number used as the basis of a scale of counting (e.g., **binary** (base 2), **decimal** (base 10), etc.). A particular base n uses the digits between 0 and n – 1; so base five uses the digits 0, 1, 2, 3 and 4. The base of a **logarithm** system is a number to which all numbers in that system are referred.

The numerator is the number on the top of a vulgar fraction, in this case 17; the denominator is the number on the bottom of a vulgar fraction, in this case 45.

$$\frac{17}{45}$$

Slide rules were traditionally used to calculate logarithmic functions. They were a mechanized form of logarithmic tables that did not require the user to know the logs. They have been all but superseded by calculators and computers.

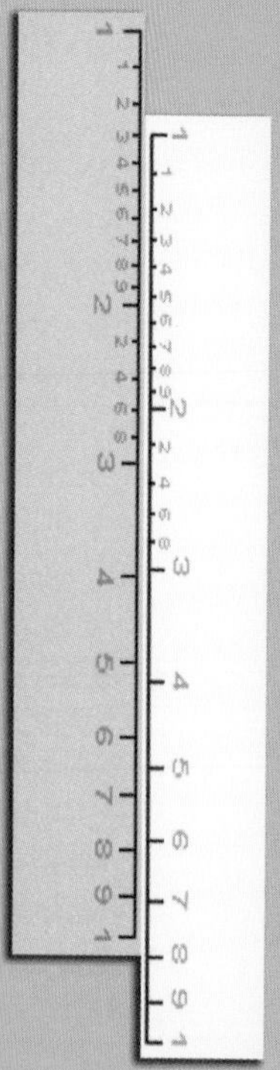

binary

base 2. The binary system of numeration uses only the digits 0 and 1. The decimal number 2 is rendered as 10 in binary; the decimal number 5 is rendered 101. The binary system is fundamental to Boolean logic and hence to the operation of modern electrical circuits and computers, because the two digits can represent two different voltages.

decimal (denary)

base 10. Decimal is the most commonly used system of numeration and uses the 10 digits 0, 1, 2, 3, 4, 5, 6, 7, 8 and 9. It is widely assumed that humans have mostly adopted the decimal system of counting because they have 10 fingers and 10 toes. Not all human civilizations have used decimal, however. Base 8, or octal, has been used by the Maya, the Babylonians and the Yuki Indians.

hexadecimal

base 16. Hexadecimal uses the following 16 digits: 0, 1, 2, 3, 4, 5, 6, 7, 8, 9, A, B, C, D, E, F. In hexadecimal, the decimal number 10 is written A; the decimal number 16 is written 10; the decimal number 100 is written 64; and the decimal number 1,000 is written 3E8. Hexadecimal is used by computers, because four **binary** numbers, or bits, can be represented easily as a single hexadecimal digit. For example, 1011 in binary is equal to B in hexadecimal.

power

the result of a number being multiplied by itself a certain number of times. Hence, 4 to the power of 6, or 4^6, is equal to $4 \times 4 \times 4 \times 4 \times 4 \times 4 = 4{,}096$. In this example, 4 is known as the **base**, and 6 is known as the exponent. The exponents 2 and 3 are used so often that they have more common names: the square and the cube respectively. Negative exponents have the effect of reciprocating the power, so 4^{-6} is equal to $1/4^6 = 1/4096$. The exponent 1 is seldom written, as $a^1 = a$. Any number raised to the power of 0 is defined as being equal to 1.

exponential

a function in which a base number is raised by a certain **power**, or exponent, to produce a given result. The inverse of an exponential function is a logarithmic function.

logarithm (log)

the power to which a base number must be raised to produce a given result. The logarithmic function is therefore the inverse of

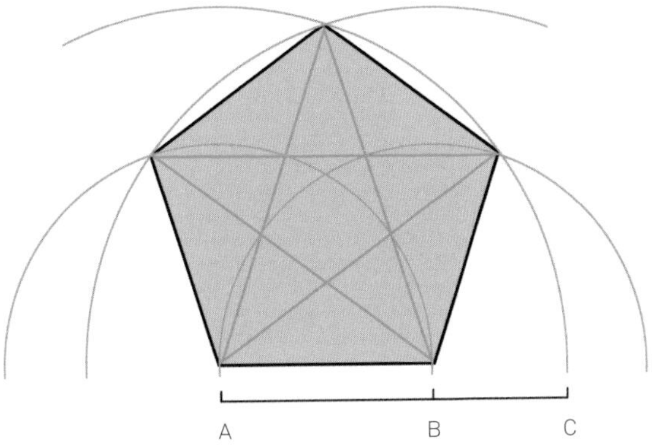

The diameter of a pentagon (the distance from one vertex to the center of the side opposite) is equal to the length of one side multiplied by the golden ratio.

the exponential function. Logarithms are represented thus: $\log x^a = b$ where $x^b = a$. So, $\log_8 64 = 2$, because $8^2 = 64$. However, $\log_4 64 = 3$ because $4^3 = 64$. Logarithms are often used to solve equations where the exponent is unknown, and they often appear in calculus, especially as the solutions to differential equations. Natural logarithms are logarithms whose base is **e**.

order of magnitude
a scale of size, typically determined by powers of 10. So, if a number b is an order of magnitude bigger than a, it is 10 times bigger.

golden ratio (ϕ)
a ratio, traditionally thought to be particularly pleasing, that occurs in mathematics, geometry and nature with some regularity. Written as a number, it is approximately equal to 1.618033, and is often represented by the Greek letter phi (ϕ). The ratio of a regular pentagon's sides to its diameter is equal to the golden ratio, and it is used to describe the **Fibonacci sequence**. Paper sizes used to be based on the golden ratio, although the current standard sizes are based on $\sqrt{2}$.

scalar
a quantity with a magnitude but no direction, in contrast to a **vector**. Distance is an example of a scalar measurement; if we say that an object is 100 meters away, we know nothing about the direction we need to travel in order to reach that object.

vector
a quantity that is described by both magnitude and direction. Vectors, therefore, are distinct from **scalars**. Velocity is an example of a vector measurement as it measures the rate and direction of motion.

Nuclear and atomic physics

quark flavors
a property by which the subatomic particles known as quarks are defined. There are six flavors of quark: up (u), down (d), charm (c), strange (s), top (t) and bottom (b). Traditional means of measuring and describing quarks are not useful, and concepts such as flavor and also color are used as ways of distinguishing one type of quark from the others.

atomic number (Z)
the number of protons found in the nucleus of an atom. All elements possess different atomic numbers, and the atomic number of an element determines its place on the **periodic table** of elements.

For this atom—which has four electrons—to be a "neutral" atom (with no overall electrical charge), the nucleus would need to contain four protons. If the nucleus contained six protons, it would be a positive ion, with a charge of +2e.

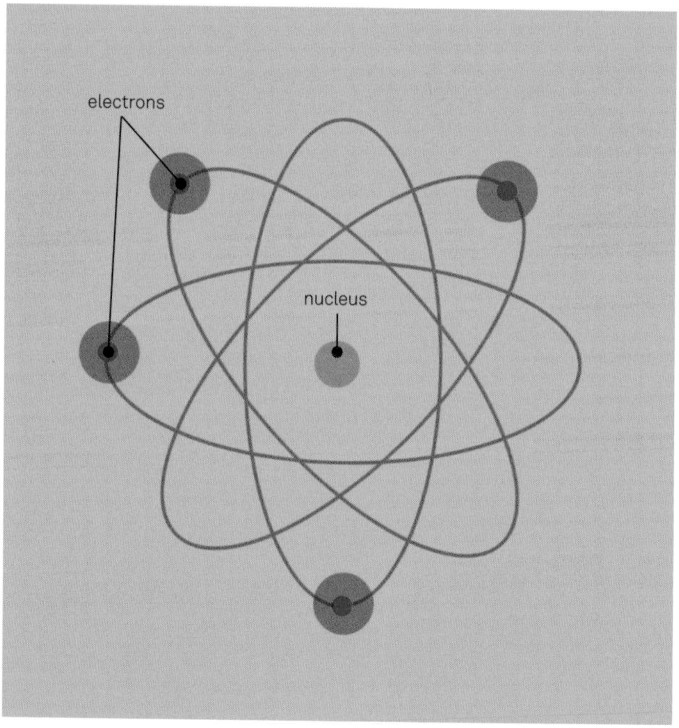

elementary charge (e; q) 👁
a unit of electric charge equivalent to the negative electrical charge carried by a single electron. It is about $1.60217646 \times 10^{-19}$ coulombs. Elementary charge was first measured by the American physicist Robert Millikan (1868–1953) in 1909. Quark particles carry charges of + ⅔ e or - ⅓ e.

electron mass
the mass of an electron. It is equivalent to approximately 9.1094×10^{-28} grams.

rest mass
in general usage, the mass of an object. The rest mass, or invariant mass, is distinct from the relativistic mass, a term used in special relativity to describe mass affected by a difference in frame of reference (relativistic mass increases for an object moves at very high speeds realtive to the observer measuring it).

de Broglie wavelength (λ) 👁
the wavelength of a particle. A wavelength is the distance between the repeated crests of a wave pattern. It is named after the French physicist Louis de Broglie (1892–1987) who proposed that any particle that has momentum also has a wavelength. The de Broglie (it is pronounced "de Broy") wavelength of a relativistic particle is equivalent to h/p, where h is the Planck constant and p is the particle's momentum.

atomic mass unit (amu)
a unit of mass used for expressing atomic and molecular masses. It is also called the unified atomic mass unit (u) or the Dalton (Da). One atomic mass unit is equal to one-twelfth of the mass of a carbon-12 atom, or 1.66×10^{-30} grams. Carbon-12 is chosen as the basis of the unit because it contains equal numbers of protons, neutrons and electrons, the principle building blocks of all atoms. Because electrons weigh much less than protons and neutrons, the atomic mass in amu is equivalent to the number of protons plus neutrons in its nucleus.Atoms of a particular element always have the same atomic number, but do not always have the same relative atomic mass.

Bohr radius 👁
the radius of the lowest energy unit in the model of the hydrogen atom described by the Dane Niels Bohr (1885–1962). The Bohr

Every object has an associated wavelength; the smaller the object, the longer its de Broglie wavelength —but even subatomic particles havevery short wavelengths.

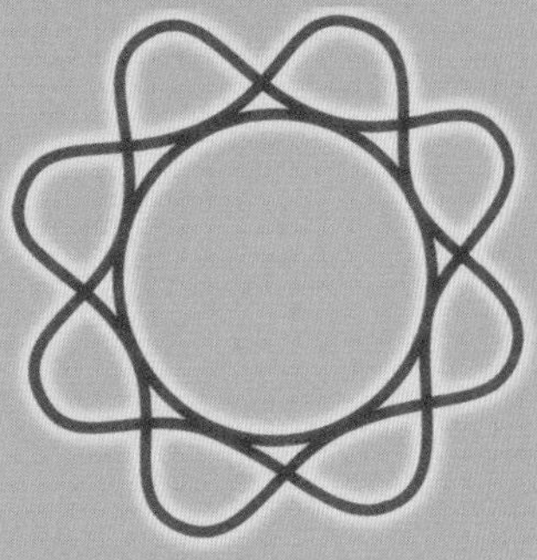

The classic representation of the hydrogen atom as shown by Niels Bohr.

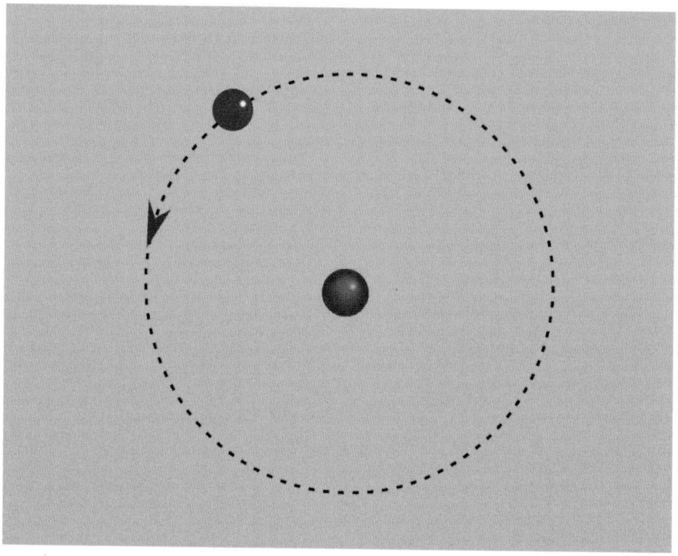

radius is used as a unit of length in atomic physics, and is equal to 5.292×10^{-11}, or about half an **angstrom**.

quantum
a unit of relative energy. The amount of energy in a quantum is proportional to the radiation it represents and is equal to the frequency multiplied by the Planck constant. An extended meaning of the word quantum is the smallest measurement into which certain other physical properties may be divided. **Elementary charge**, for example, is the quantum of electrical charge.

becquerel (Bq)
the SI unit of radioactivity, named after French physicist Antoine-Henri Becquerel (1852–1908), who shared the 1903 Nobel Prize for Physics with Marie and Pierre Curie for discovering natural radioactivity in uranium salts. One becquerel is equal to the radiation caused by the radioactive disintegration of 1 atomic nucleus in 1 second. It is approximately equal to 27 picocuries (**curie**).

curie (Ci)
a unit of radioactivity. Originally defined as the radioactivity of 1 gram of radium-226, it was agreed in 1953 that 1 curie should be equal to 3.7×10^{10} atomic disintegrations per second. Named after Marie and Pierre Curie (1867–1934 and 1859–1906 respectively).

sievert (Sv)
an SI unit measuring the effective dose of radiation received by a living organism. It is named after Swedish physicist Rolf Sievert (1896–1966). The sievert is equal to the actual dose of radiation multiplied by a factor which varies according to how dangerous the radiation is—the higher the factor, the more dangerous the radiation. The effective dose of natural background radiation is approximately 1.5 millisieverts a year. If the effective dose is raised between 2 and 5 sieverts it can cause hair loss, nausea and, in some cases, death. One sievert can be subdivided into 100 rem.

gray (Gy)
an SI unit of energy measuring the absorbed dose of radiation. It is named after English radiobiologist Louis Gray (1905–65). One gray is defined as being the dose of 1 joule of energy absorbed by 1 kg of matter. It is equal to 100 **rad**. The gray differs from the **sievert** in that it does not differentiate between more and less harmful forms of radiation.

half-life
the time taken for half the atoms in a sample of a particular radiuoactive isotope to decay. To reduce the radioactivity of the original isotope to 0.1 percent of its original level takes just under 10 half-lives, Half-lives vary wildly between elements. The half-life of uranium-238 is 4,500 million years; that of radium-221 is 30 seconds, and many unstable isotopes have half-lives measured in fractions of a second. However note that measurements of radioactive risk are complicated by the fact that the decay products of the original isotope may themselves be radioactive.

half-value layer
the amount of shielding required to reduce the intensity of radiation by half. The half-value layer is a means of measuring the intensity of a radiation source. Different radiation sources have their half-value layers defined in terms of different materials. For example, the half-value layers of X-rays are generally described in terms of aluminum or copper thicknesses.

k factor
a measure of the strength of the gamma rays produced by a radio-active material. The k factor is measured in **roentgens** per hour at a distance of 1 cm from a source with a disintegration rate of 3.7×10^7 per second (or 1 millicurie).

A dosimeter is usually shaped like a pen so that it can be easily clipped to the clothes. It can then measure the wearer's exposure to harmful environmental factors such as radiation or noise.

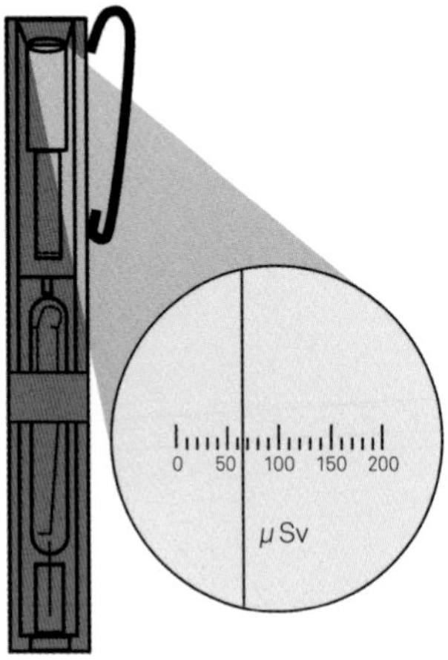

roentgen (R)

a unit of ionizing radiation. It is named after German physicist Wilhelm Roentgen (1845–1923), who discovered the X-ray. As radiation hits an atom, it ionizes it by removing one or more electrons. This causes many of the biological effects of radiation, and the roentgen is a means of measuring these effects. One roentgen is the dose of radiation that liberates positive and negative charges of 2.58×10^{-4} coulombs of electric charge per kilogram of air.

radiation absorbed dose (rad)

a metric unit of radiation dosage, equal to 0.01 **gray**. In other words, 1 rad is equal to 0.01 joules of energy absorbed per kilogram of tissue.

rutherford (Rd)

a unit of radioactivity, named after New Zealand physicist Ernest Rutherford (1871–1937) One rutherford is equal to 1 mega-becquerel (**becquerel**), or one million radioactive disintegrations per second.

Hounsfield scale

a scale for measuring radiodensity, or how transparent a substance is to X-rays. A radiolucent substance is more transparent to X-ray photons than a radiodense substance, and so has a lower score on the Hounsfield scale, affecting its appearance in X-ray images and CAT scans.

dosimeter

a device used to measure exposure to a potentially harmful environment. A radiation dosimeter measures exposure to ionizing radiation. Because radiation absorption is cumulative, the dosimeter should be worn every time the individual is in the presence of a radioactive source.

Geiger counter

a device for measuring levels of radioactivity. Named after its creator, German physicist Hans Wilhelm Geiger (1882–1945), it detects and counts ionizing particles in the atmosphere. The Geiger counter can detect photons, alpha radiation, beta radiation and gamma radiation, but not protons. Some Geiger counters use a visual measuring device such as a needle; others produce an audible click. The Geiger counter has now been largely superseded by a new version of the device known as a halogen counter, which operates with a much lower voltage and has a longer life.

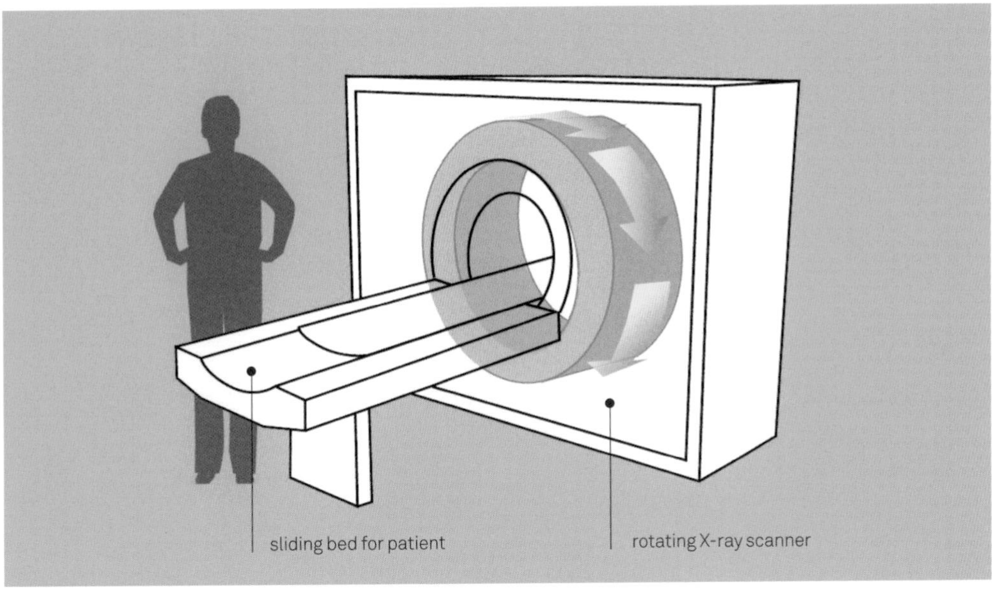

A CAT scanner uses the radiodensity scale devised by Godfrey Hounsfield to measure the internal anatomy of the body.

background radiation
radiation occurs naturally in the environment, derived from natural sources within the Earth, in the atmosphere and in surrounding space. A small part of the background radiation measured as weak microwaves from space originates from the Big Bang itself and is known as the cosmic microwave background. Background radiation varies, but it is generally in the range of 0.3–0.4 rem (see **sievert**).

critical mass
the minimum amount of fissionable material that will sustain a nuclear chain reaction. Various factors affect the critical mass of a material, including its nuclear properties, its shape, its purity and whether the material is surrounded by a neutron reflector that bounces escaping particles back into it and helps sustain the chain reaction. The shape with the minimum critical mass is a sphere. The critical mass for a sphere of uranium-235 without a reflector, for example, is 50 kg, compared to 15 kg for uranium-233.

kiloton (kton)
a unit of energy equivalent to that released by the explosion of 1,000 tons of Trinitrotoluene (T.N.T.), roughly equal to 4.18×10^{12} joules. One thousand kilotons are a megaton. The kiloton is used to describe the destructive power of nuclear weapons—the bomb that was dropped on Hiroshima, for instance, produced about 15 kilotons of energy, while the largest nuclear weapon ever detonated produced 57 megatons of energy.

Energy

Wood is a relatively poor fuel, in terms of energy released per kilogram, but has the great advantage that it is easy to grow more of it.

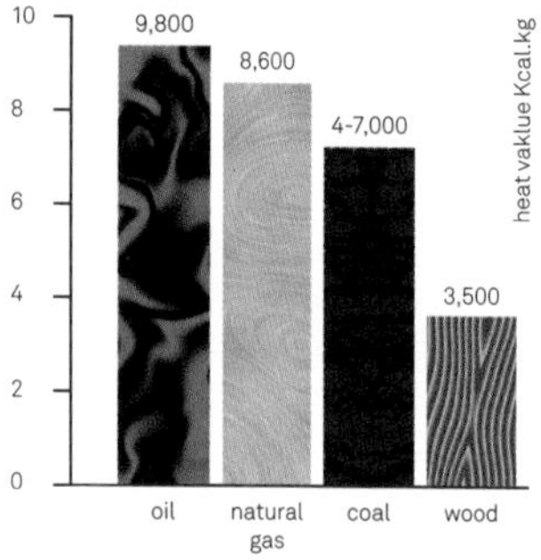

energy

the measure of the ability of an object (whether a photon of light, a tennis ball or an entire galaxy) to affect another object. It is often defined as the object's capacity for work, but this definition fails to take account of **entropy** (heat energy unavailable to perform work). Energy can take a wide variety of forms, though all can be classified as either kinetic (related to movement) or potential (stored—as, for example, in a chemical bond, coiled spring, or an object raised above the ground). Energy can be converted from one form to another, but can never be destroyed or created.

calorie (cal)

a unit of energy, defined as the amount of energy needed to raise the temperature of 1 gram of water by 1°C at a pressure of 1 atmosphere. Nutritional **Calories** (with a capital C) are 1,000 calories (with a small c), and are technically called kilogram-calories (although kilo calories is also acceptable) as they are the amount of energy needed to heat 1 kg of water by 1°C. As with the **Btu**, the temperature at which a calorie is calculated affects the precise energy value.

joule (J)

the SI unit of energy and work. It is the energy used to exert a force of 1 **newton** over a distance of 1 meter (and the work performed by doing so); the energy to move an electrical charge of 1 coulomb over a potential difference of 1 volt; or the energy to produce 1 watt of power for 1 second. One J is equal to 0.24 calories or just under $^1/_{1000}$ of a **Btu**.

kilowatt-hour (kWh)

the unit used by power companies to measure the amount of electricity used by a household. One kilowatt-hour is the amount of electricity that a 1 kW device uses in 1 hour (or a 2 kW device in half an hour), and is equal to exactly 3,600,000 J. Sometimes called a "unit" of electricity.

chemical energy
usually used simply to mean the **bond energy** of a chemical compound. It is occasionally used to describe the amount of energy released (or absorbed) during a chemical reaction, which is simply the total energy of all bonds created minus the total energy of all bonds destroyed, but this should more properly be termed the reaction energy.

heat of combustion
the energy released as heat when a material undergoes complete combustion—i.e., the substance being burnt reacts with as much oxygen as possible. It can be measured per mole, per unit mass, or per unit volume of the material, and is usually used for comparing fuels (the higher the value, the better the fuel). A substance with a negative heat of combustion actually absorbs heat as it burns.

calorimeter
a device used to measure the energy emitted (usually as heat) or absorbed by a chemical reaction or state-change (such as melting). The two most common kinds of calorimeter are the "bomb" calorimeter, used to measure energy changes in fast reactions, and the differential scanning calorimeter, which measures the change in energy over time, but many other specialist designs exist for various purposes.

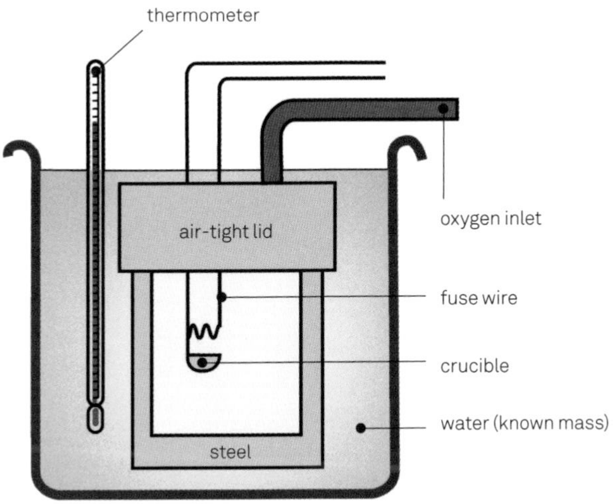

A "bomb" calorimeter, for calculating the energy released by an explosive reaction or burning substance.

Energy can be stored in a variety of ways. While capacitors store electrical energy, batteries store chemical potential energy, which is only converted to electrical energy when they are connected to a circuit.

electrical

elastic

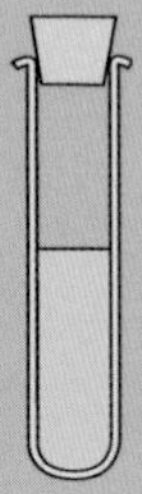

chemical

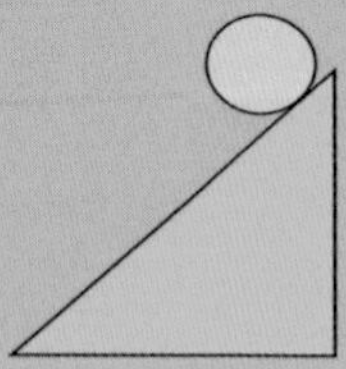

Missing?

mass–energy conversion factor

according to Einstein's theory of special relativity, the mass of a body is a measure of the total energy contained within it (i.e., mass is simply another form of energy). This is the famous $E = mc^2$ where E is the energy equivalent to the rest mass m (the mass of the object when it is completely stationary). The energy released in nuclear reactions is due to this conversion; the products of fission or fusion have a slightly lower mass than the reactants.

surface energy

the energy (usually measured in joules per square meter) required to break the chemical bonds within a substance and create a new surface. As there is an interaction between the atoms on each side of the surface, the actual value of the surface energy depends on both the substance and its surroundings—the surface energy between diamond and air is not the same as that between diamond and water. Generally, the greater the surface energy, the harder it is to break an object (though other factors may also come into play).

potential energy (PE) 👁

in effect, stored energy which an object can release to perform work (by moving to a state with lower energy). Types include gravitational PE (an object at a height can fall—PE = the mass of the object × the distance it can fall × the strength of the gravitational field), elastic PE (stretched or compressed materials can spring back), and electrical PE (which drives electrons round a circuit). Chemical **bond energy** is also a form of potential energy.

kinetic energy

the energy possessed by an object as a result of its movement (including heat energy), equal to half the mass multiplied by the square of the velocity. Since energy cannot be destroyed, this is equal to both the energy needed to accelerate the object to its current velocity, and the energy needed to stop it again.

force

the rate of application of energy to change the velocity or shape of an object. Forces always exist in equal and opposite action-reaction pairs—a football hitting a wall has a force exerted on it by the wall that brings it to a halt, but it exerts an equal force on the wall (and thus Earth). The total change (equal to the force multiplied by the time it was applied for) is sometimes called the impulse.

newton (N)
the SI unit of force. A newton is the amount of force needed to accelerate a mass of 1 kg by 1 m s^{-2}.

dyne (dyn)
an outdated small unit of force, not in general use. A dyne is the force required to accelerate a mass of 1 gram by 1 cm per second per second. One hundred-thousand dyne = 1 newton.

pound-force (lb-f)
the force exerted by the Earth's gravity on a 1-lb weight, equal to just under 32 lb.ft.s^{-2} (**pound feet** per second squared) or about 4.4 newtons.

thrust
the creation of a force by pushing material backwards relative to the desired direction of movement, leading to forward movement as a result of the equal and opposite reaction force. The greater the ratio of the thrust to the object's weight, the greater the acceleration produced. Thrust is usually measured in either pounds or newtons. The thrust produced is equal to the mass of the material pushed backward multiplied by the acceleration given to it.

Propellers, wheels and people all generate thrust by pushing directly against their surroundings. A rocket's thrust is generated by creating a large amount of hot gas then accelerating that in the opposite direction to the rocket's movement.

Propeller thrust

Propeller thrust

Rocket thrust

Swimming thrust

A pair of meshed gears can be used either to increase torque (lowering speed) or increase speed of rotation (lowering torque).

dynamometer
a machine to measure the power (and torque) produced by an engine (of any kind), or the power input required by a piece of machinery.

gearing
the use of gears (or belts and pulleys) to transfer force from one place to another, usually changing the speed of rotation (and the torque) in the process. Note that a gearing system cannot change the amount of power (which is always slightly less coming out than going in, as a result of frictional losses), but can increase torque by decreasing speed of rotation or vice versa.

moment
a measure of the turning effect caused by a force around a pivot, equal to the force multiplied by the distance from the pivot. Moment has the same dimensions as energy, but is measured in different units (for the SI system, newton-meters). A moment of 1 newton-meter uses 1 joule of energy to move through an angle of 1 radian. Torque is another word for moment, applied particularly to engines and motors.

power
the amount of energy used (or work done) per unit time.

watt (W)
the SI unit of power, equal to 1 joule per second (or, in electrical units, a current of 1 ampere across a **potential difference** of 1 volt). For many purposes, power consumption of modern equipment is often stated in kilowatts, and commercial power generation rarely deals with quantities smaller than a megawatt.

efficiency
in energy (and heat transfer) terms, the efficiency of a machine or process is the proportion of the energy put in thatis used to do useful work, stated either as a fraction or a percentage. A notional perfectly efficient machine (with an efficiency of one) would require no energy to function, and waste no energy to its surroundings as unwanted heat —but such a machine is impossible in practice. Internal combustion engines tend to have an efficiency of below 20 percent, while steam turbines in power stations are around 35 percent efficient.

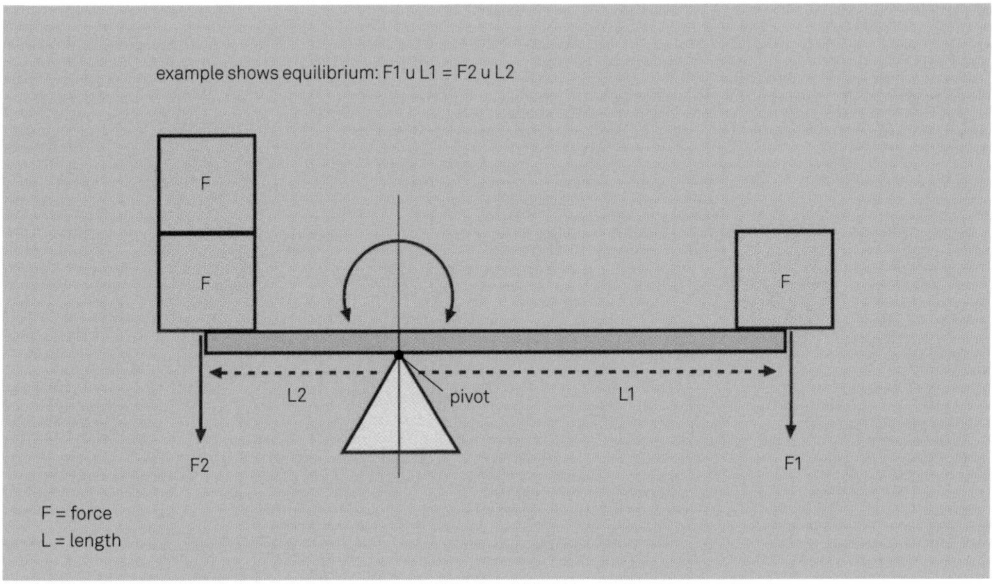

Equal and opposite moments generated by non-equal forces cancel out around a pivot.

horsepower (hp)
a very old unit of power, still used as the measure of power for internal combustion engines. Originally defined by the scientist James Watt as the average rate of work of a typical horse walking on a treadmill to power machinery, and later numerically as 550 **foot pounds** per second (746 W).

work
the energy transferred between objects by applying a force over a distance, equal to the **scalar** product of the force and movement (both of which are **vectors**). If the movement is directly along the line in which the force is applied, this is simply the magnitude of the force multiplied by the distance moved. Work has the same units as (and is essentially a special case of) energy.

Speed and flow

speed
the rate of an object's motion, equal to distance traveled per unit time. Speed is a **scalar** quantity: 30 mph in any direction is 30 mph.

velocity
the rate at which an object's location changes with time. It is the **vector** equivalent of speed, and measures both the rate and the direction of the object's movement. The magnitude of an object's velocity is its speed.

A geostationary satellite in circular motion is an example of continuous acceleration while maintaining a constant speed.

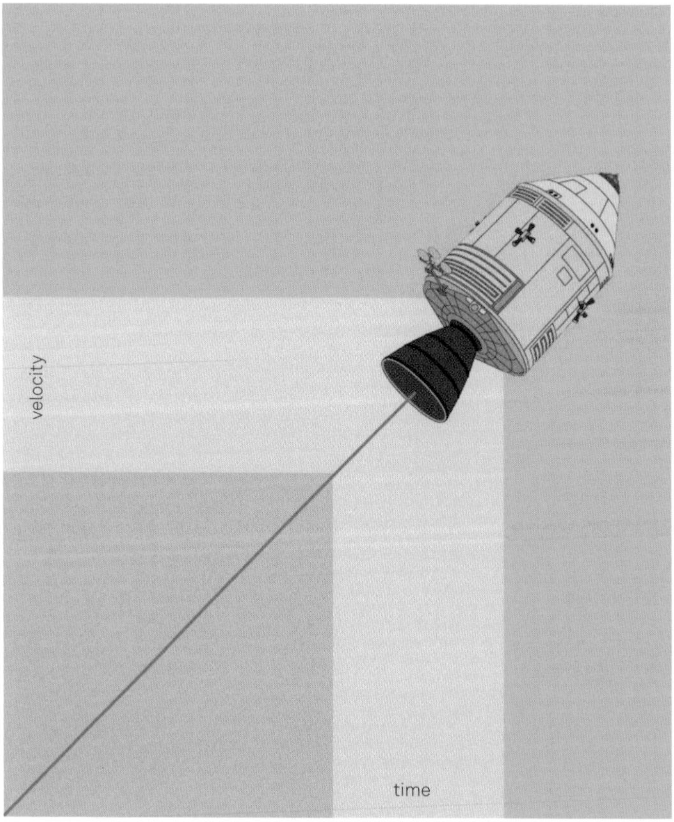

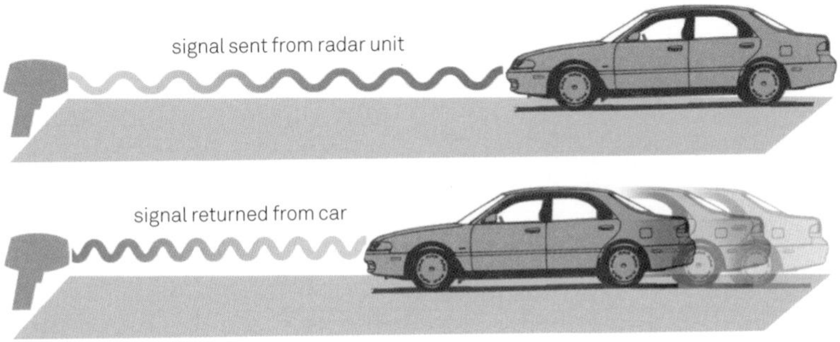

The faster a car moves toward you, the more the Doppler effect compresses the sound waves and the higher the pitch of the sound you hear.

acceleration

the rate at which an object's **velocity** or **speed** changes with time, measured in change of speed per unit time (in SI units, meters per second squared; ms^{-2}). For velocity, a **vector** acceleration is required, while change of speed is **scalar**. Negative acceleration in terms of speed is often called deceleration but the term is not normally used with changes in velocity. Where a vector acceleration is not exactly parallel to the original velocity, the direction of movement changes, usually as well as the speed.

meters per second ($m\ s^{-1}$)

the SI unit of **speed** (and **velocity**). One $m\ s^{-1}$ is equal to 3.6 km/h, or a little over 2 mph. Used for most scientific work that involves speed or velocity.

Doppler effect (Doppler shift)

the apparent change in the frequency (and wavelength) of a wave caused by the movement of the observer relative to the source of the waves. The effect is the same for a given relative velocity even if either the source or observer is stationary. For example, to a listener standing still on the ground, as a plane flies overhead the sound of the engine appears to change as its velocity changes. The engine's apparent frequency increases as the plane approaches and decreases as it flies away. The phenomenon was first proposed by Austrian mathematician Christian Doppler (1803–53).

knots

the unit of speed for ships (and one of those used for aeroplanes). One knot is 1 nautical mile per hour (about 1.15 "land" miles per hour, or 0.51 meters per second). The **air speed** of an aircraft is usually measured in knots, while the ground speed is usually recorded in either kilometers or miles per hour.

Appearances can be deceptive—the moon orbits Earth once a month, but the Earth's rotation means that the moon is only visible from one place for a few hours—rather than days—at a time.

angular velocity

the velocity of rotation of a spinning object. Since it is necessary to specify a direction (the axis around which the object is spinning, the poles of which remain stationary) this unit is a **vector**. It can also be used to describe the speed at which one object is orbiting another, although this becomes complicated if the object being orbited is also rotating. Angular velocity is measured in angle per unit time; the SI unit is radians per second.

revolutions per minute (R.P.M.)

a unit of rotational speed. A rotational speed of one R.P.M. is equal to an **angular velocity** of 6° per second.

momentum

a measure of how difficult it is to stop a moving object. Momentum (equal to **mass** times **velocity**) is not the same as **kinetic energy** (half of mass multiplied by the square of speed)—a 10 kg rocket moving at 1,000 mph and a 1 tonne car moving at 10 mph have identical momentum but the kinetic energy of the rocket is a hundred times higher. In addition to the difference in magnitude, momentum is a **vector**, while kinetic energy is **scalar**. So if two identical objects collide while traveling in opposite directions at the same speed and both stop, their combined momentum (which was zero before they hit) remains the same, but their kinetic energy changes drastically, being converted to other forms of energy, such as sound, heat, and deformation of the objects' internal structures.

muzzle velocity

the speed of a bullet or other projectile as it leaves the barrel of a gun or other launcher. Since there is no directional information (the bullet is always moving along the length of the barrel unless something is badly wrong), this is strictly a speed, not a velocity.

Mach number

the speed of an object relative to its surroundings (e.g., the **air speed** of an aircraft) expressed as a multiple of the speed of sound. An object with a Mach number greater than 1 is termed supersonic; when the Mach number reaches 5 it becomes hypersonic.

air speed

the speed of an aircraft or other vehicle relative to the surrounding air. Since the air, especially at the altitudes used by modern jet planes, is usually moving relative to the ground, air speed is almost never the same as the vehicle's ground speed. An aircraft with a cruising speed of 500 mph and a 50 mph tailwind will have an effective ground speed of 550 mph.

ground speed
the speed of a vehicle, usually a plane or ship, relative to the ground. Ships also have a water (or sea) speed, which is equivalent to an plane's **air speed**.

terminal velocity
the speed at which the air resistance to downward movement of a falling object is exactly equal to the gravitational force acting on it. Terminal velocity for a falling person (with parachute closed) ranges from about 120 mph with limbs spread to over 200 mph in a streamlined "diving" position.

speed of light (c)
the speed of light in a vacuum, exactly 299,792,458 meters per second (or approximately 670 million mph), which the laws of physics say can never be exceeded. Any observer will measure the value of c relative to themselves as being the same, regardless of their **velocity** relative to the source of the light. This leads to a variety of extraordinary effects, of which the apparent distortion of time and distance experienced by a person or object traveling at speeds close to c is probably the best known.

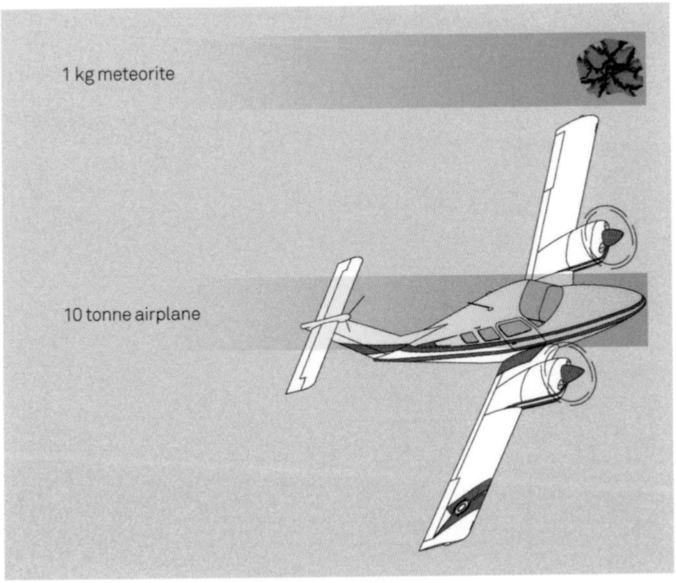

A 1 kg meteorite moving at 10,000 mph has the same momentum as a 10 tonne airplane taxi-ing at 1 mph on the ground. To have the same momentum as a 10 tonne airplane flying at 150 mph, the meteorite would need to travel at 1,500,000 mph.

critical velocity

the minimum speed relative to its surroundings at which a liquid or gas will display turbulence. Below this speed, the flow is "smooth," and turbulence is prevented by the fluid's **viscosity**.

viscosity (dynamic viscosity)

a measure of the resistance of a fluid (a liquid or gas) to flow, both internally and through what may be called "fluid friction." (Although solids can flow, albeit very slowly, this is an entirely different process—see **creep**—and solids do not strictly have a viscosity.) Viscosity is independent of pressure, but does change with temperature—as the temperature increases, the viscosity of a gas will rise, while that of a liquid tends to fall.

kinematic viscosity

the dynamic **viscosity** of a fluid divided by its density. This can be a more useful measure of a fluid's behavior than dynamic viscosity alone, especially where flow under gravity is involved. If two fluids (e.g., honey and machine oil) have the same dynamic viscosity but different densities, they will behave differently—and their kinematic viscosity reflects this difference.

In theory, people on two airplanes traveling in opposite directions will see time pass at a different speed on the other airplane to the passage of time on their own. In practice, such effects are only significant at very high fractions of the speed of light.

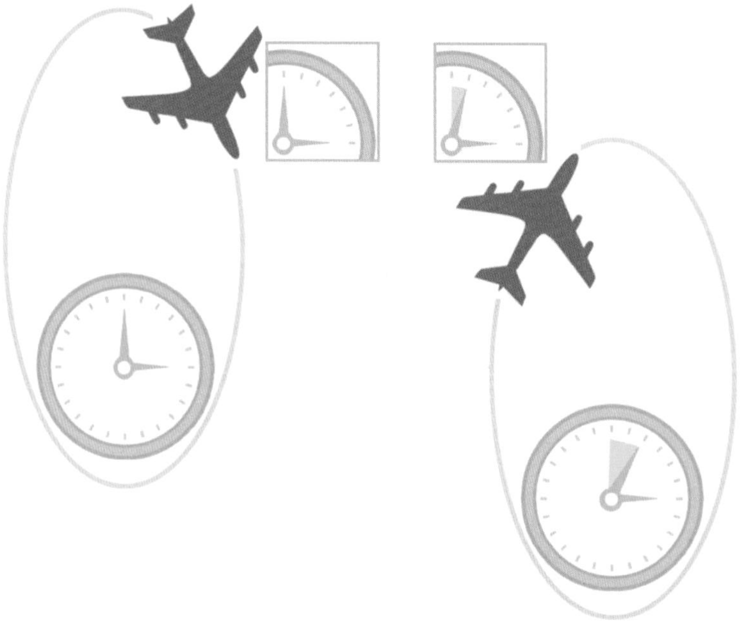

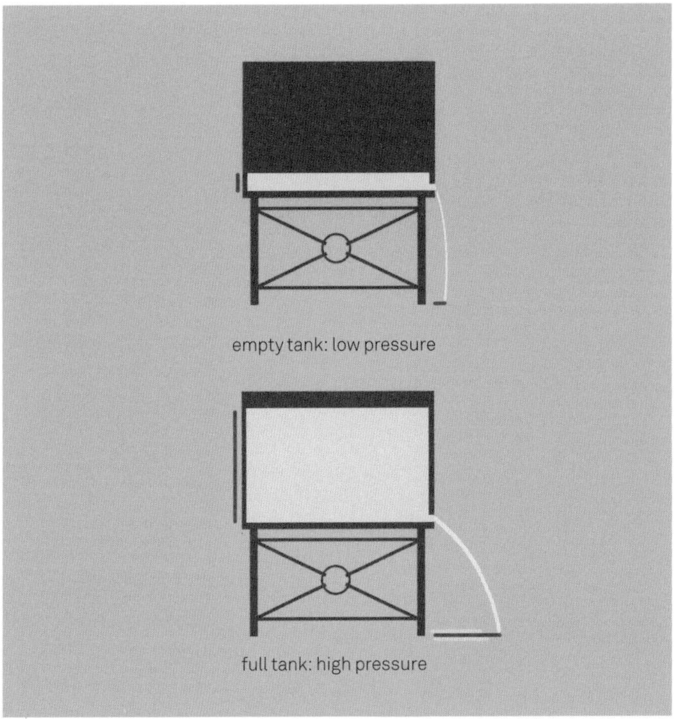

The higher the pressure, the faster water flows through a standard-sized hole.

viscosity grade (V.G.)
one of several measures of viscosity, usually used for lubricants and oils. **S.A.E.** viscosity grades for engine oil were originally a measure of the length of time a fixed quantity of oil took to flow through a test orifice at 100°C, but current I.S.O. (and A.S.A.) viscosity grades are equal to the kinematic viscosity of the fluid at 40°C. The viscosity index measures how much a fluid changes with temperature.

cusec
a unit of flow rate; 1 cubic foot per second. The metric equivalent is a cumec, equivalent to 1 cubic meter per second (just over 37 cusec).

gallons per minute (gal min^{-1})
a small unit of flow rate; 1 U.S. gallon per minute is only 0.0022 cusec, making 1 cusec equal to 448 gallons per minute.

megaliters per hour (ML/hr)
a measurement of flow rate used in industrial processes, water supply and the study of rivers. It equals one million litres per hour, or about 10¼ cubic feet per second. For major rivers, or very large-scale industrial measurements, flow will sometimes be specified in megaliters per minute, or even per second.

Mass and weight

Because of the hugely different sizes and densities of the planets of the solar system, their gravitational forces are correspondingly different. So, a mass of 1 kg would feel like only 166 grams on the moon, and 2.364 kg on Jupiter.

1	Jupiter	2.364
2	Mercury	0.378
3	Saturn	1.064
4	Venus	0.907
5	Uranus	0.88
6	Earth	1
7	Neptune	1.125
8	Moon	0.166
9	Pluto	0.067
10	Mars	0.377

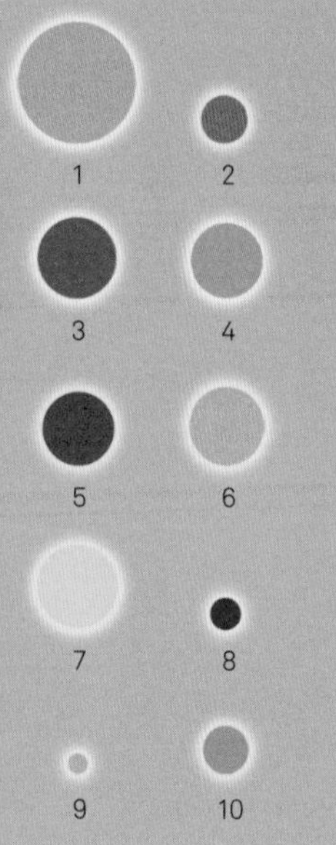

weight

the gravitational force acting on a body, measured in **newtons** or **dynes** in the SI system, or **pound-force** in the U.K. imperial system. In many scientific contexts, it is essential to distinguish between weight and mass, and strictly speaking the base units of the metric and imperial systems (kilograms and pounds) are units of mass; it is, however, quite usual to come across these units loosely used to mean the mass under the force of the Earth's gravity at its surface—in fact the imperial system defines pounds as weight under this assumption. The difference becomes immediately apparent when comparing the weight of a body on Earth, and then on the moon; although its mass is unchanged, its weight is considerably less due to the moon's weaker gravity.

mass

the amount of matter in a body, the inertial mass being a measure of its resistance to change of motion, and the gravitational mass a measure of its attraction to another body. The base units of mass in use today are mainly the kilogram in the SI system, and the pound in the U.S. Customary and U.K. imperial systems.

gravity

the force of attraction between bodies due to their mass. This force was first defined by Isaac Newton (1642–1727) in his Law of Universal Gravitation, which states: "Any two particles of matter attract one another with a force directly proportional to the product of their masses and inversely proportional to the distance between them." This is expressed in the equation, $F = G (m_1 m_2/d^2)$, where F is the gravitational attraction between bodies m_1 and m_2, and d is the distance between them; G is the gravitational constant, equal to 6.6732×10^{-11} n m^2 kg^{-2}.

g

the symbol for acceleration due to gravity of a body in free fall. It is also used loosely as a unit of acceleration, where 1 *g* is equal to 9.80665 meters per second squared (about 32.17405 feet per second per second) at the Earth's surface, but this varies in practice with altitude and latitude.

density (ρ)
the ratio of the **mass** of a body to its volume. Density is expressed in terms of mass per unit volume, such as kilograms per cubic meter, or pounds per cubic foot.

center of gravity
the point at which the resultant of gravitational forces experienced by all the particles of a body acts. In a uniform gravitational field, it is the same as the center of **mass**: the point where the mass of the body may be considered to be concentrated for ease of calculations.

troy weight system
an ancient system of weights that survives to some extent in the jewelery trade in the U.K. and North America.

avoirdupois weight system
the system of weights in use in Britain from the 14th century to the introduction of metrication in the 1960s, and widely used in the English-speaking world up to the present day. The base unit of the system is the **pound** (lb). The term derives from the old French *avoir du pois* (meaning "having weight").

apothecaries' weight system
a system of weights originally used by 17th-century apothecaries for measuring very small amounts. It differs from the troy weight system in its subdivisions of the troy ounce, known in this system as the apothecaries' ounce. The apothecaries' ounce is divided into **drams** (or drachms), **scruples** and **grains**.

tonne (t)
an SI unit of mass, equal to 1,000 kilograms. To distinguish it from the U.K. imperial **ton** (to which it is very close in value) and the U.S. ton, it is also sometimes called a metric tonne or metric ton.

ton
a unit of mass or weight in the U.K. imperial and the U.S. Customary systems, but with different values in each system. The ton in common usage in the U.K., also known as the long ton, is equal to 2,240 lb (1,016.0416 kg), whereas the U.S. ton, or short

U.K. imperial system

Avoirdupois:	
1 ounce (oz) = 16 drams	= 438.5 grains
1 pound (lb) = 256 drams = 16 ounces	= 7000 grains
1 stone	= 14 pounds
1 quarter	= 2 stones
1 cental	= 100 pounds
1 hundredweight (cwt)	= 112 pounds
1 ton (or long ton)	= 2240 pounds

Troy:	
1 pennyweight	= 24 grains
1 ounce troy (oz tr)	= 480 grains = 20 pennyweights
1 pound troy (lb tr)	= 12 ounces troy

Apothecaries:	
1 pound apoth.	= 12 ounces apoth.
1 ounce apoth.	= 24 scruples
1 drachm (in US dram)	= 3 scruples
1 scruple	= 20 grains

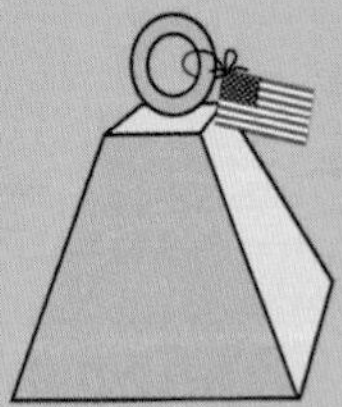

U.S. Customary system

Avoirdupois:	
1 dram	= 27.34 grains
1 ounce (oz)	= 16 drams
1 pound (lb)	= 16 ounces
1 hundredweight (or short hundredweight)	= 100 pounds
1 ton (or short ton)	= 2000 pounds

Troy:	
1 pennyweight	= 24 grains
1 ounce troy (oz t)	= 20 pennyweights
1 pound troy (lb t)	= 12 ounces troy

ton, is equal to 2,000 lb (907.18474 kg). Historically, the ton was also used as a unit of volume, especially in shipping dry goods.

hundredweight (cwt)
a unit of mass or weight in the U.K. imperial and U.S. Customary systems, but, like the ton, of which it is a subdivision, with different values in each system. In Britain and many other English-speaking countries, it is equal to 112 lb, or 1⁄20 of a (long) ton (50.80208 kg), but recent usage in North America has produced a unit known as a short hundredweight, 1⁄20 of a short ton (100 lb or 45.359237 kg).

stone
a mainly British unit of weight, equal to 14 lb. Nowadays its use is restricted almost entirely to descriptions of body weight, but even there is disappearing in favor of the kilogram.

kilogram (kg)
the base unit of mass of the SI system, one of the seven base units from which all other units are derived. It is specifically a unit of mass, rather than weight or force, unlike the base units of traditional systems where these were effectively interchangeable. Until 2019, the definitive kilogram was a cylinder of platinum-iridium alloy kept by the Bureau International des Poids et Mesures in Sèvres, France. However the unit is now defined in terms of the second, metre and several fundamental constants of nature.

Despite metrication, body weight is still normally expressed in stone and pounds in the U.K.

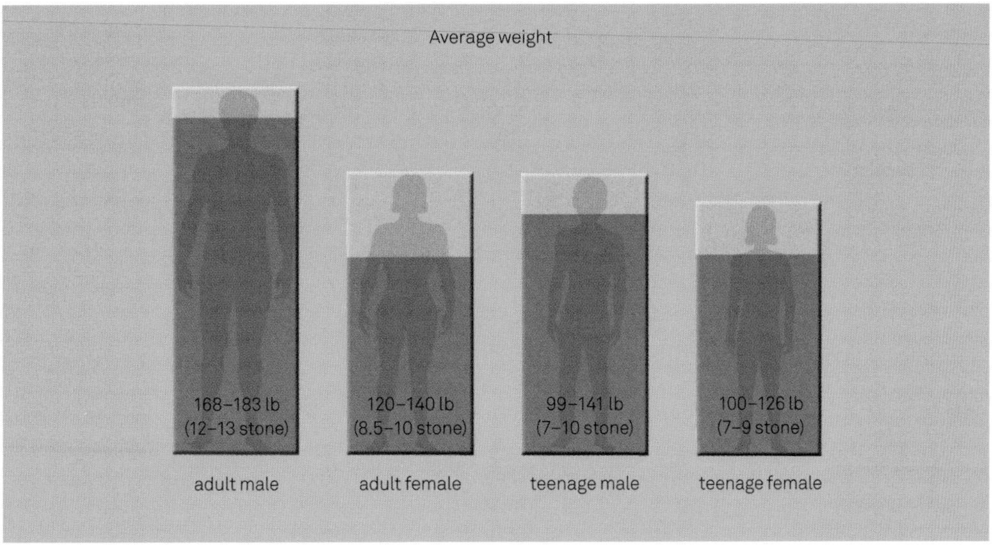

The shekel was a common unit of weight throughout the ancient Middle East, but is probably better known today as the coin of the same weight used by the Hebrews.

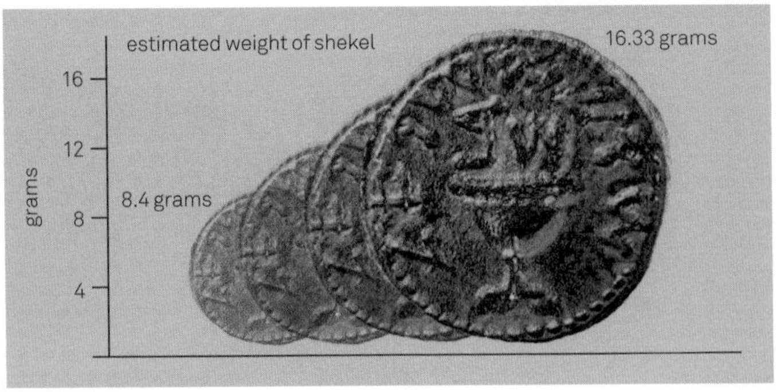

pound (lb)

the base unit of mass in the U.K. imperial and U.S. Customary systems, equivalent to 0.45359237 kg. The use of a pound in one form or another dates back to ancient Rome, where the *libra pondo* (pound of weight) was commonly used, and from which derive many traditional European units of weight; it is also the source of the abbreviation lb, short for *libra*. The U.K. and U.S. **avoirdupois** pound is divided into 16 ounces, unlike the Roman **libra** and similar units used in southern Europe which divided into 12. The mostly obsolete Troy and apothecaries' pounds are identical in value, and equal $^{144}/_{175}$ avoirdupois pounds, or 0.373.242 kg. An additional distinction is made by the abbreviation lbf, for **pound force**, which is a unit of force rather than mass: 1 pound force is equal to 4.448221615 newtons.

libra

a unit of weight in ancient Rome, still in use informally in many Mediterranean and Spanish-speaking countries. Like the name of the constellation and the horoscope sign, the Latin word *libra* originally referred to the balance scales used for measuring the weight. The unit of weight had a variable value, but came to refer specifically to around 0.722 U.K. pounds (0.32745 kg), which was divided into 12 *unciae*. In France it is known as the *livre*, and has come to mean exactly 500 grams.

shekel

an ancient Babylonian unit of weight used widely in the Middle East, and adopted by the ancient Hebrews. Its exact size is disputed, but values range from around 8 to 16 grams. Some sources confidently give an exact equivalent of 252 grains (about 16.33 grams), others give 8.4 grams: it is probably safer to assume that we do not know. The shekel was also a Hebrew coin of the same weight.

penny-weight
a unit of weight in the troy system, equal to 24 grains, or 1⁄20 of a troy ounce (1.552 grams).

gram (g)
a small unit of mass in the SI system, at one time the base unit of mass of the metric system known as the C.G.S. (centimeter-gram-second) system, but now defined as 1⁄1000 of the International Prototype Kilogram. The unit derives from the Greek *gramma*, which was similar in value to the Roman scruple. The original French spelling *gramme* is still seen occasionally, but is not internationally recognized.

grain
a small unit of weight in the U.K. imperial and U.S. Customary systems. Its name derives from an early definition as equivalent to the weight of one grain of wheat or one barleycorn. It was in effect the original base weight of the British systems, as it was used to define the pound in each usage: one **avoirdupois** pound = 7,000 grains, one **troy** pound = 5760 grains. The term grain is also used in the jewelery trade as equal to one-quarter of a **karat** (50 milli-grams); in this context it is sometimes called the pearl grain after its use for measuring the weight of pearls.

point
a small unit of mass used for measuring gemstones, equal to one-tenth of a **karat** (20 milligrams).

Technology and leisure

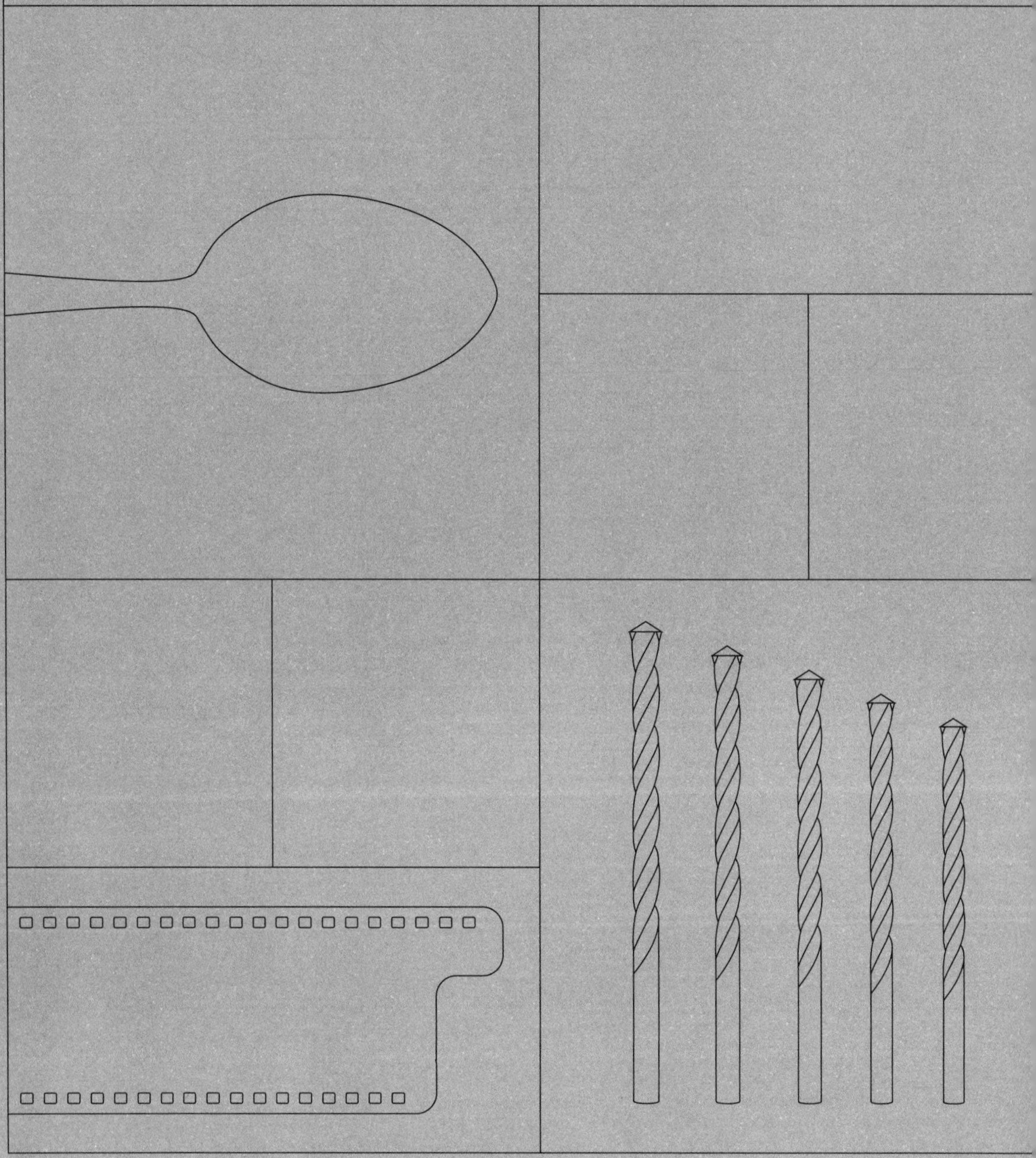

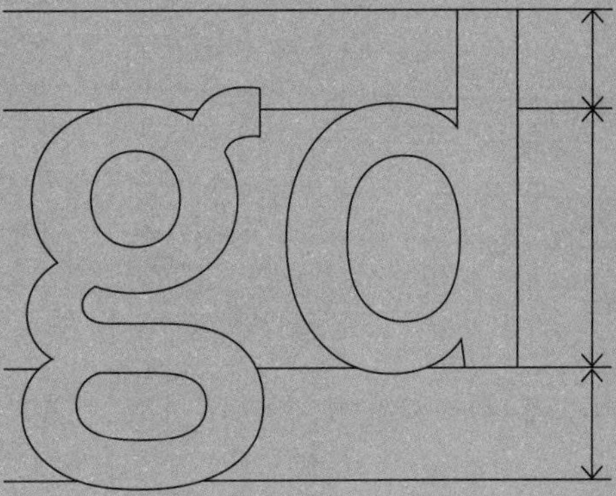
gd

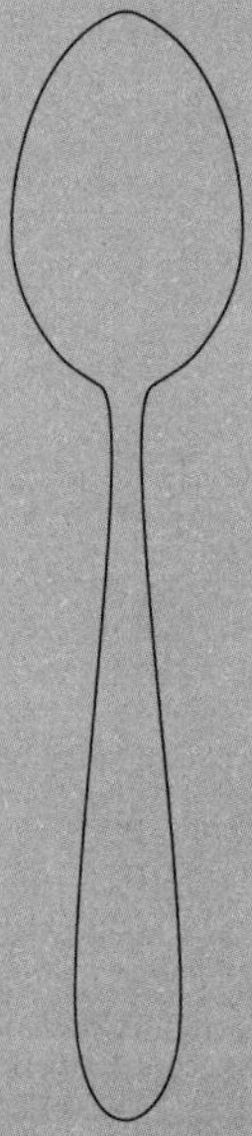

Computers and communications

kibi-

an uncommon but accurate prefix used to solve a conflict in computer terminology. The problem arises because many computer components (e.g., memory) expand more logically by powers of two than powers of 10. Thus, while the kilo- prefix is usually taken to mean 1,000, a kilobyte of memory is the nearest equivalent in powers of 2: that is, 2^{10} or 1,024 bytes. The difference between round decimals and the nearest power of 2 increases for larger multiples, leading to potentially confusing and conflicting descriptions of capacity and power for different components. The International Electrotechnical Commission (I.E.C.) therefore suggested in 1998 that a set of new prefixes be defined in powers of 2 (kibi- = 2^{10}, mebi- = 2^{20}, gibi- = 2^{30}, etc.) to allow distinction between the two sizes, but these have not been widely adopted.

bit (b)

the smallest unit of computer memory, a bit is a single one or zero in the binary (base 2) counting system. It has become ubiquitous because values in this system can be clearly represented by just two different and easily distinguished levels of electric charge, current, laser light or many other properties. Most modern computers deal with bits in larger groups such as **bytes**, but "embedded" high-tech systems with smaller memories (such as factory robots) may use the value of individual bits to keep track of simple on/off conditions.

Big-endian (starting with the largest digit) and little-endian (starting with the smallest digit) numbers both have their uses, but will use the same arrangement of bits to indicate very different numbers.

binary number	**decimal equivalent** (big-endian)	**decimal equivalent** (little-endian)
1000	8	1
0001	1	8
00000001	1	128
10010001	145	137
1110	14	7
111110	62	31
011111	31	62
1111	15	15

endianness

The direction from which an electronic component interprets a binary number consisting of several bits. Systems can be either big-endian (with the leftmost digit indicating the largest power of 2, in the same way as normal decimal numbers) or little-endian (starting with the digit indicating the smallest power).

byte (B)

a byte is a group of eight bits, recorded (and acted on) as a group. It therefore has 256 (2^8) possible values, which in most cases are used to represent the decimal numbers either from zero to 255, or from −127 to +128.

nybble (nibble)

a nybble is a binary string of four bits or half a byte, representing 16 possible values.

word

the unit of memory dealt with internally by a computer's central processing unit (C.P.U.), and passed between C.P.U. and the main memory. The size of either or both can vary between different types of computer, but most modern PCs are built around processors that use a 64-bit word.

floating point

a floating point number is actually two numbers stored together, each occupying a fixed number of bits specified by the computer and programming language used. These two numbers represent a decimal integer and a power of two by which that integer is multiplied to give the floating point value. While this system allows a very wide range of numbers to be stored in a single relatively compact format, it results in the number stored being only an approximation. This can cause problems when adding small numbers to large ones, or where the last few digits of a number are the important part, as a floating point system may be unable to distinguish between 1,000,000,000 and 1,000,000,004.

character

a single letter, digit, punctuation mark or similar. Characters are represented in modern computing systems using developments from a code system known as ASCII (American Standard Code for Information Interchange). First developed for use with teleprinters that sent and received signals as electric pulses over telegraph wires, the original ASCII system used seven bits (128 possible

The table shows how floating point works using a 3-digit decimal number (i.e., from −999 to +999) and a power of ten in the range +3 to −3. As you can see, in this somewhat extreme example there is no way to store a number such as 4,002 (or 4.002).

integer	power of 10	no. stored
42	0	42.0
42	+2	4200
42	-2	0.42
426	-2	4.26
-426	-2	-4.26
426	+1	4260
427	+1	4270
-427	+1	-4270
427	+3	427000
427	-3	0.427
4	-3	0.004
4	+3	4000

The resolution used increases from left to right, with each step to the right making the letter much crisper, hence easier to read.

1 8 d.p.i.
2 16 d.p.i.
3 32 d.p.i.
4 64 d.p.i.
5 128 d.p.i.
6 256 d.p.i.

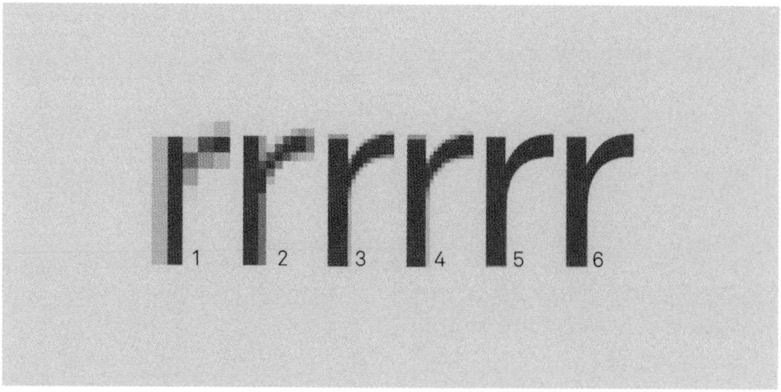

values) to represent the upper- and lower-case letters of the English alphabet, the digits 0 to 9 and various punctuation marks (with 33 non-printing values used as control codes for the machine). Since the 1990s, ASCII itself has been superseded by the more flexible Unicode system, which uses two or more bytes to incorporate a vast range of characters from other languages, accents and other symbols.

string
a sequence of characters "strung" together; the usual way of storing text in a computer program. The term is sometimes also used (with an appropriate modifier word) for connected sequences of other memory units, e.g., a "binary string" is a string of bits.

resolution
the amount of detail present in a picture or displayed by a screen. The higher the resolution, the better the image will tend to look, and the more it can be magnified before appearing blocky. A computer monitor has a "native" resolution, as the display surface is composed of a large number of pixels, but the monitor will also be able to display other resolutions if necessary. However, these other resolutions may either appear blurred or have jagged edges (especially on LCD monitors), as pixels cannot be assigned partially to one colour and partially to another.

D.P.I. (Dots Per Inch)
the standard measure of resolution for scanners and printed images. Exactly as you expect, the number of spots along a 1-inch line at which the scanner records the color, or the printer places a dot of ink. Since this is a linear measure and images cover an area, doubling the resolution in D.P.I. will quadruple the size of a scanned image file (or reduce the printed image to quarter-size).

pixel
the unit of display in the memory of a computer (or other digital system) or on a video screen—the name is a contraction of "picture element."

voxel 👁
the 3D equivalent of a pixel, widely used for representation of scientific data (e.g., medical imaging, such as M.R.I. scans) and, less often, for certain types of computer game.

benchmark/specmark
a standard test (or series of tests) designed to measure the speed of a computing system (or one of its components). Benchmarking allows comparisons to be made between different items designed to perform the same task. S.P.E.C. (the Standards Performance Evaluation Corporation) benchmarks are designed carefully to simulate real usage.

MIPS (Millions of Instructions Per Second)
a badly-flawed measure of computer C.P.U. speed using integers. Different processor designs use different numbers of instructions to perform the same task, and the speed will often be calculated by the manufacturer using a program and language selected or even designed to produce a high result.

FLOPS (Floating point Operations Per Second)
the floating point equivalent of MIPS, and only a little more useful as a genuine indicator of performance. The speed of super-computers is often quoted in teraflops (billions of FLOPS) – for example Sony's 2013 PlayStation 4 games console had a peak performance of 1.84 TFLOPS, while the 2020 PlayStation 5 can archive 10.28 TFLOPS.

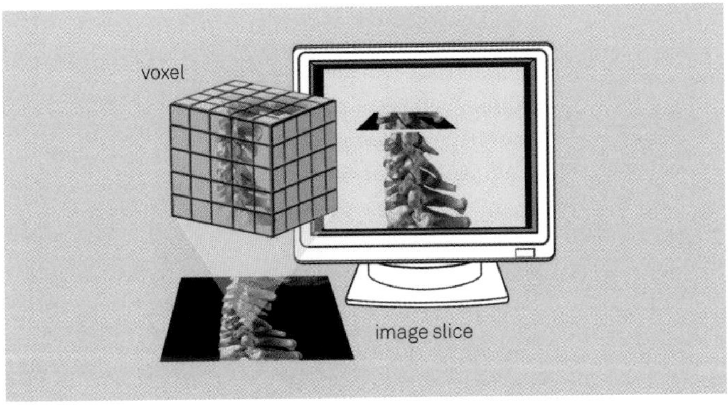

The storage of 3D information as voxels allows medical imagers to examine any cross-section through it at any angle.

Radio spectrum divisions

VLF 3–30 kHz
radio navigation, time signals and similar simple broadcasts

LF (long wave) 30–300 kHz
aircraft navigation, some AM (amplitude modulation) broadcast radio (in Europe), amateur radio (Europe) and the U.S. "lost band" for experimental use

MF (medium wave) 300–3000 kHz
most AM radio is in this band

Medium wave 500–1700 kHz
U.S. and Canada AM radio

Short wave 3–30 MHz
a wide range of uses, including international broadcasting, amateur radio (U.S.) and "numbers" stations for communicating with spies

VHF 30–300 MHz
FM radio, T.V. stations, 2-way radio and the VOR aircraft navigation system

UHF 300–3000 MHz
mobile phone networks and TV broadcast, including HDTV

Microwave 3–300 GHz
radar, satellite communications, wireless computer networks

baud

the base unit of bandwidth (speed of data transfer) for digital communications, equal to one bit per second. Low transmission speeds are also sometimes quoted in bytes (or characters) per second, while higher rates (such as cable and fibre modems) are normally given in megabits per second, which give more impressive-looking numbers than the equivalent in bytes (1 megabit, for example, is only 125 kilobytes).

bit rate

a measure of the rate at which data is transferred or processed that is effectively equivalent to baud, but is applied to a wider range of digital technologies (such as television signals, hard drive speeds, and the reading and writing of Blu-ray and other optical disc technologies, and has therefore somewhat superseded the older term even in telecommunications. The 4G (fourth-generation) standard for cellular phone networks can achieve peak download data rates of around 1Gbit/s, while the newer 5G is anticipated to be 10 times faster once networks are fully deployed. For comparison 4K and Ultra-HD Blu-ray players deliver around 100 Gbit/s of data to the television via an HDMI cable. Streaming 4K movies, meanwhile, use compression algorithms to reduce the required bitrate to around 16Mbit/s or less.

bandwidth 👁

in telecommunications, a range of frequencies assigned to a single purpose. Medium wave radio, for example, usually has a bandwidth of 10 kHz, with 5 kHz empty bands between channels (signals are carried by modifying the amplitude or strength of the radio waves). In digital applications, a technique called multiplexing is used to transmit multiple streams of data at different frequencies within the available bandwidth.

erlang

a unit of telephone traffic. One erlang is equivalent to one person using the line continuously (one hour of conversation per hour). If a line has enough bandwidth for 10 simultaneous conversations, and half that bandwidth is being used, the traffic is five erlangs.

attenuation

the amount by which a signal (including radio, electrical currents, laser light and earthquake shockwaves) is reduced in power as it travels from one place to another, normally measured in decibels per unit distance. Attenuation is a problem in modern communications networks (though less so for glass optical fibers than for traditional copper cables), and such networks usually need to have "repeaters" to amplify the signal at regular intervals.

frequency response

the accuracy of a system's reproduction of input at a specific frequency; theoretically this applies to any equipment, but it is generally only used for electronics (and most often for sound reproduction). Typically quoted as a range of frequencies and a variation in decibels, meaning that the object will reproduce any signal in that range to within the specified variation.

radio frequency bands

the "radio spectrum" (which includes all electro-magnetic wavelengths longer than infrared) is divided into a number of bands, each of which is allocated to specific uses.

REN (Ringer Equivalency Number)

a measure of the amount of power required to make a telephone (or other device) ring. Typically, the maximum total REN for all the equipment plugged into a normal domestic phone line is around four or five; if the total is above this, the phones may not ring at all.

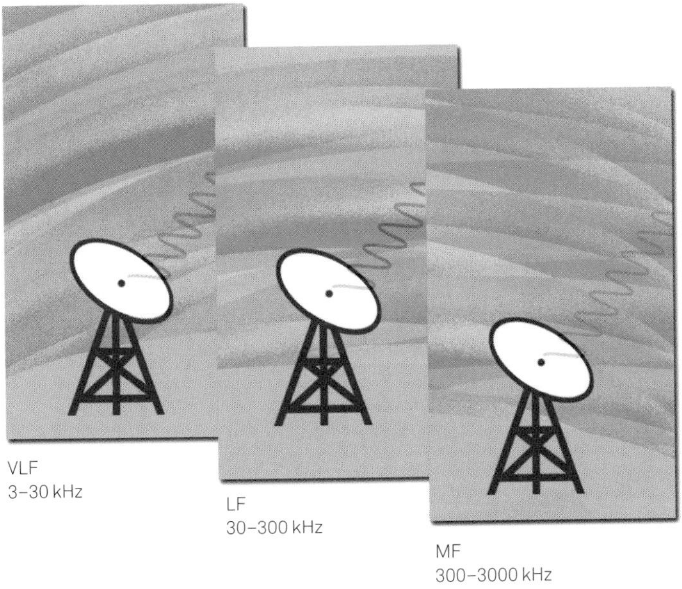

Among other things, radio antennae can detect the microwave background radiation "left over" from the start of the universe.

Engineering

strength

the ability of a material to withstand **stress**, measured in force per unit area. Sometimes also applied to structures, in which case it is the ability of the structure to cope with load (an external force applied to a structure).

hardness

the resistance of a material to indentation, scratching or abrasion. It is measured by several different methods. Rockwell tests use a cone-shaped diamond (or "brale"), hard steel balls or other items to penetrate the material using a load of 10 kg, followed by a larger load of up to 150 kg, and the extent of penetration is measured. Brinell tests use a hard steel or carbide ball with a much higher load than Rockwell tests— as much as 3,000 kg—for a specific time period, which varies depending on the material. The Brinell number is load divided by the indentation area in square millimeters.

toughness

the ability of a material to withstand impact, dependent upon the distribution within the material of the stress and strain caused by the impact. The opposite of toughness is brittleness. The toughness of steel and other metals may be measured by the **Charpy impact test**.

Using the Rockwell hardness test here, the permanent depth of indentation is the amount of penetration of the specimen after the removal of the large load, while the initial small load remains.

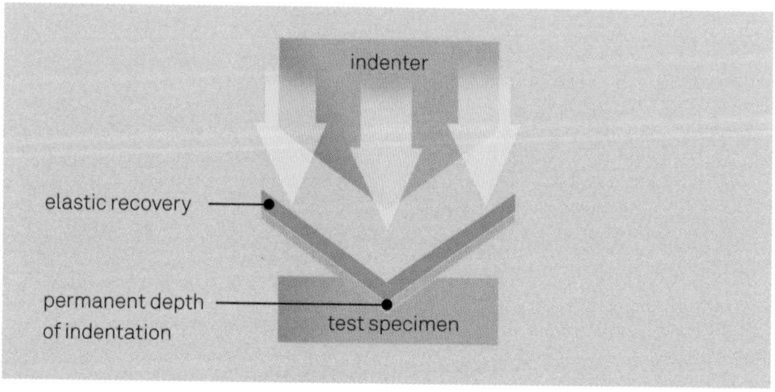

stress
an applied force or system of forces that tends to strain or deform a body. It is defined as force per unit area, and can be expressed as kilograms per square millimeter. Stress takes a number of forms, notably shear stress (acting parallel to the surface, causing a change in shape, without particular volume change) and tensile stress (causing increase in volume or length, or both). The SI unit of stress is the **pascal**; in the U.S., pounds per square inch are often used.

strain
the amount of deformation caused by **stress**, measured as the ratio between the change in the length of an object after stress was applied and its length beforehand. When a sample is being tested for its reaction to tensile stress, it changes while the strain is being applied, so a more accurate measure, "true strain," has to be used. This involves taking "instantaneous" measurements and adding them together.

pascal
the SI unit of pressure or stress, equivalent to one newton per square meter (1.45 × 10^{-4} pounds per square inch).

standard atmospheric pressure
an arbitrary representative value for atmospheric pressure at sea level, originally calculated as the pressure exerted by a column of mercury 760 mm tall at 0°C, but now defined as one atmosphere or 101.325 kilopascals.

tension
the force exerted on an object or material by stretching it, causing it to increase in volume or length, or both. Applying tension until the substance breaks is known as tensile testing.

compression 👁
the opposite of **tension**, caused by "squeezing" an object or material, and reducing its volume. For metals, results of compression tests are usually the same as those for tensile tests, but for other materials, such as polymers, this is not the case. The term compression has a particular application with regard to the internal combustion engine, and refers to the compressing of the air/fuel mixture before combustion. The compression ratio is the ratio between the largest volume in a cylinder and the smallest, when the piston has reached maximum compression.

The compression ratio in an internal combustion engine: the ratio between the volume in a cylinder at maximum (top) and minimum compression (bottom).

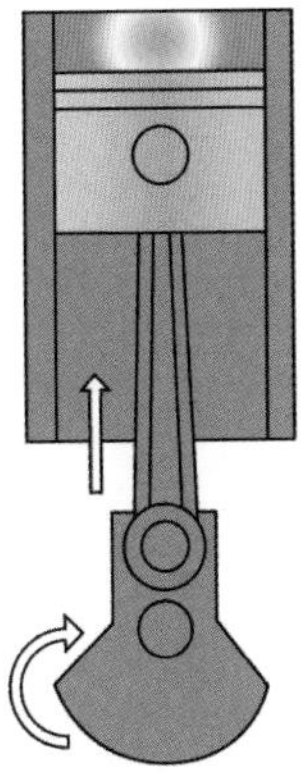

compression stroke

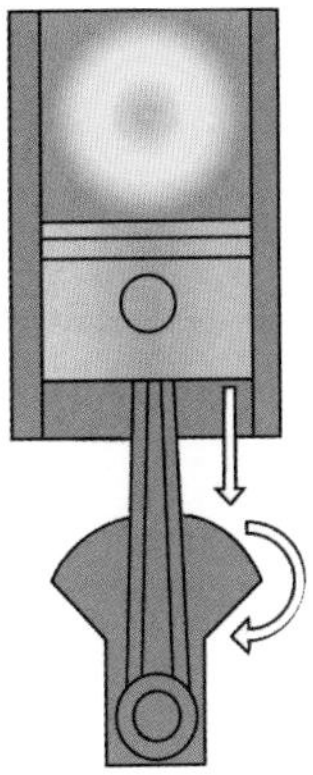

exhaust stroke

elasticity (elastic deformation)

the extent of the ability of materials to resume their original shape after stresses of **tension** or **compression** have been applied to them. Materials such as rubber have a high degree of elasticity, and can return to their original shape very easily, but even rubber will eventually deform, beyond its **elastic limit**.

hysteresis

the degree to which a strain depends on the history of previous stresses as well as the present stress. If a stress is only partially removed, most materials will show a greater strain than if the remaining stress had been applied to the unstrained object. When the stress is completely removed the material may—or may not—return to its original shape. Hysteresis also applies in magnetism; when some objects are placed within a magnetic field and then removed, some residual magnetism will remain in the object.

elastic limit

the limit to which an elastic material can be stretched by **stress** without causing a permanent change in its shape.

elastic modulus

the ratio of **stress** to **strain** under particular conditions. This usually remains constant as some materials that have several distinct elastic phases until the **elastic limit** is reached. Young's modulus of elasticity refers to tensile testing. It is the applied stretching force per unit area

The number allocated to each material on the Mohs hardness scale does not represent any relationship with the others on the list.

- 10 diamond
- 9 corundum
- 8 topaz
- 7 quartz
- 6 feldspar
- 5 apatite
- 4 fluorspar
- 3 calcite
- 2 gypsum
- 1 talc

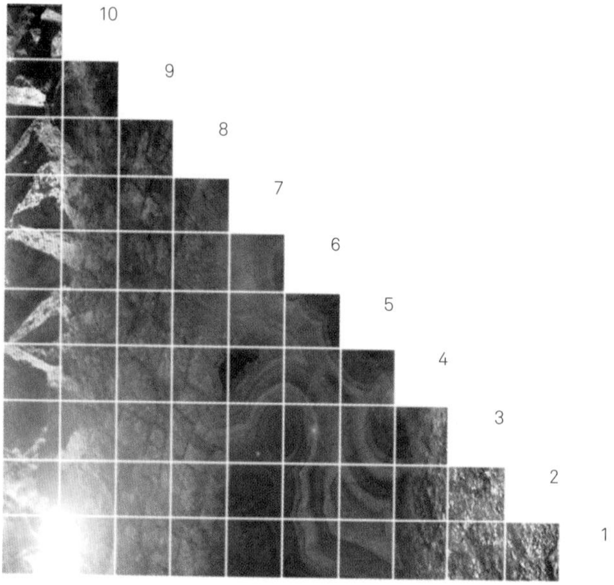

divided by the increase in length per unit length. The shear modulus of **elasticity** is the shearing force per unit area divided by the angle of twist. The bulk modulus of elasticity is the compressive force per unit area divided by the change in volume per unit volume.

ultimate tensile strength
the point at which a material or an object under tensile stress actually breaks. There are corresponding terms, "ultimate tensile stress" and "ultimate tensile strain," which are the values of **stress** and **strain** at the point of breaking.

creep
the slow deformation of material or an object under constant stress. There are normally three stages of creep: "primary creep," during which the strain or deformation increases rapidly with time; "secondary creep," with a slow increase in strain over time; and "tertiary creep," when the rate of creep increases again and breaking point is eventually reached.

Griffith crack length
the limit of the length of a crack in a material that can be tolerated before catastrophic failure occurs. In a large construction, the longer the crack length the better, because longer cracks are easier to spot and areas of potential danger can therefore be eliminated.

Mohs hardness scale
A rather crude but practical method of comparing the hardness or resistance to scratching of minerals, invented by German mineralogist Friedrich Mohs (1773–1839). It is not a scale as such, but a list of 10 materials in order of hardness. Each mineral is tested by scratching another material on the list: if it can scratch it, it is harder; if it is itself scratched by the material, then it is softer.

Vickers hardness test
a test for **hardness** using a square-based, pyramid-shaped diamond that is pressed into a metal. The force is applied using a load that can vary between 1 kg and 100 kg, and it is applied for 10 to 15 seconds. The size of the impression made is an indication of the metal's hardness—the smaller the impression, the harder the metal.

Shore hardness test
a test of the hardness of rubber and plastic materials. It uses a hardened indenter which is pressed into the material and the depth of penetration is then measured.

The Shore hardness test uses a Durometer. The scale shows the amount of penetration of the material.

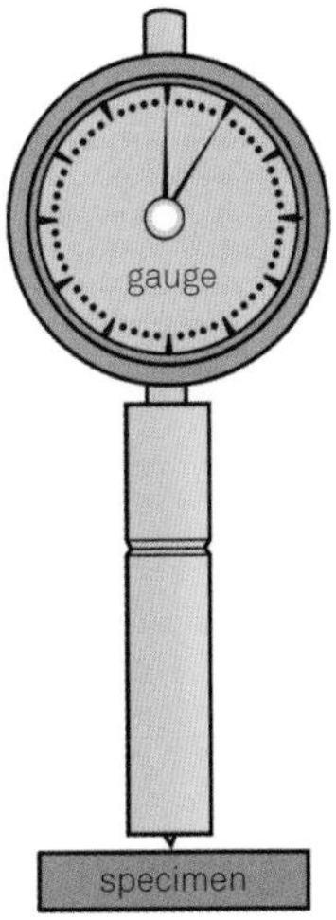

In this 3 pt bend test, the amount of bend is measured by the gauge at the central point of pressure.

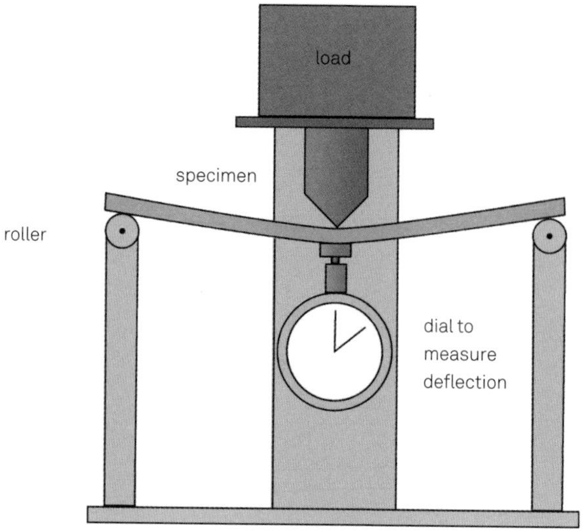

three-point bending apparatus

Charpy impact test
a rudimentary test for impact toughness, still used because it is an economical indicator useful in quality control decisions. It is generally used for assessing the toughness of metals, but similar tests are used for other materials. In the test, a pendulum is used to break by one blow a piece of metal, notched in the middle and supported at each end. The energy absorbed is measured in joules, and this absorbed energy represents the impact strength of the material. The similar Izod test also uses a pendulum but the sample is clamped at the bottom, and the pendulum strikes the top.

bend test
a test usually performed on metal to determine whether it has enough ductility to bend without breaking. It can be used on sheet, strip, plate or wire, and different criteria apply in each case, but generally a standard specimen is bent through a specified arc and, in the case of strip, the direction of grain flow is noted and whether the bend is with or across the grain.

friction
the force between two materials in contact with each other impairing their ability to slide or roll against each other. The coefficient of friction is a measure of this force, and is very low for two materials that slide together easily, and very high for two

that strongly resist movement against each other. The size of the area of contact between the two materials does not matter.

tribometer
a device for measuring the force of resistance between two materials sliding or rolling against each other. In other words, a friction tester.

S.A.E. oil grades
the various grades of lubricating oils. The system was devised by the American Society of Automotive Engineers (S.A.E.), and classifies oils in terms of their viscosity. A light oil is 10 S.A.E. and a heavy oil 40 S.A.E.

pounds per square inch (psi)
the imperial or Customary (**avoirdupois**) measure of pressure or stress. One psi equals 6.895 kilopascals. This measure is often still used in the U.S.

base box
a unit measuring the thickness of tinplate or another form of metallic coating, like galvanizing. It is equivalent to the area formed by 112 sheets of metal, 14 × 20 inches, or 31,360 square inches. The weight in pounds of one base box is the base weight.

tire sizes
the markings on the tire wall, giving information about the size of the tire and the wheel. These are now standardized, and the main information is in the form 215/65 × 15, where 215 is the size of the width of the tire in millimeters, 65 is the aspect ratio (the height to width ratio) and 15 is the size of the wheel in inches.

drill sizes
there are a huge number of drill sizes, in both metric and imperial systems. The imperial system has a series of numbers from 1 to 80, beginning at 1 for a size of 0.2280 inch, and getting progressively smaller, with 80 being 0.0135 inch. Larger sizes have letters, beginning at A for 0.2340 inch. Metric sizes are in millimeters.

The huge array of drill sizes is necessary because of the need for precision in some applications.

1 ½ = .500
2 ¼ = .250
3 ⅛ = .125
4 1⁄16 = .0625
5 1⁄32 = 0.313

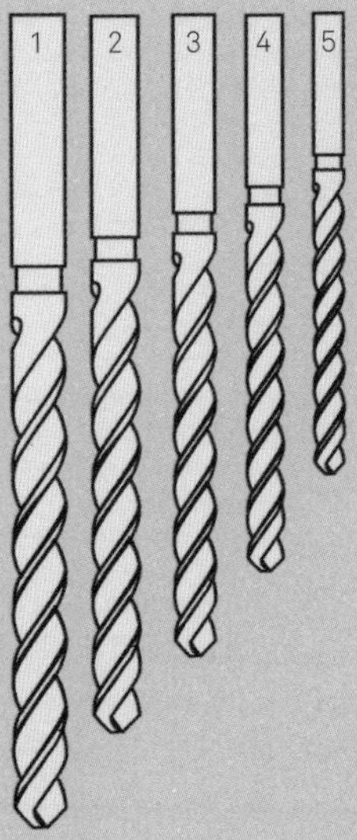

Finance, coins and currencies

Principal currencies of the world

Country	Currency
U.S.A.	dollar, cent
Canada	Canadian dollar, cent
Australia	Australian dollar, cent
New Zealand	New Zealand dollar, cent
U.K.	pound sterling, penny
France*	euro, cent(ime)
Germany*	euro, cent
Italy*	euro, cent
Spain*	euro, cent
Russia	rouble, kopek
China	yuan, jiao, fen
Japan	yen
India	rupee, paisa
Indonesia	rupiah
South Korea	won
Israel	new shekel, agora
South Africa	rand, cent
Brazil	real, centavo
Argentina	peso, centavo

**Note: several other countries in Europe also use the euro, and many plan to*

pre-decimal British currency

prior to the introduction of decimalization in 1971, the British currency, the pound, was divided into 20 shillings (denoted by the initial "s"), and each shilling was divided into 12 pence (denoted by the initial "d," from the Latin *denarius*). There had also been further subdivisions of the penny, into halfpennies and farthings (a quarter of a penny). Notes were issued for 10 shillings, 1 pound, 5 pounds, 10 pounds and higher denominations. Historically, a gold 1-pound coin, known as a sovereign, had also been issued. Another former coin, the guinea, had a value of 1 pound and 1 shilling.

Roman coins

Roman coins were issued for use throughout the vast Roman Empire, and some remained in use well into the Empire's decline. They were also used outside the Empire as a useful means of exchange. The emperor Augustus standardized coinage, minted in seven denominations (namely the *aureus, denarius, sestertius, dupondius, as, semis* and *quadrans*), using four different metals—gold, silver, brass and copper.

Greek coins

unlike Roman coinage, ancient Greek coins were not standardized because the various city states issued their own currency, in different designs. The gold *stater* was one basic unit. There were silver *staters*, and often the name *stater* was given to the chief (silver) *drachm* coin issued in a particular state. *Drachm* coins came in various multiples, e.g., the *tetradrachm* was 4 *drachma*. *Drachma* were too large in value to be practical. Similarly, the *obol* was too large in value (and too small in size) to be of much use in everyday transactions.

cash

the original low-value Chinese coin, with a square hole in the center, which first appeared in the fourth century B.C.E. and was regularly issued over thousands of years. Larger denominations of cash were also produced: 2, 5 and 10 cash. The Chinese name for the coin is the *tsien*. The word also refers to Indian and Indonesian low-value coins, though there is some dispute about a link with the English word "cash," meaning "ready money."

krugerrand

a South African coin used for investment only, first issued in 1967. It contains one troy ounce of gold. It was named for Paul Kruger, first president of the Republic of South Africa, whose head appears on the coins.

louis d'or

often shortened to simply "louis" (and sometimes known internationally as a "pistole"), a former French gold coin worth 10 *livres*. Originally issued in 1640 in the reign of King Louis XIII, it continued to be struck until the Revolution of 1789.

napoleon

a gold coin with 20 francs, originally issued in the time of Napoléon Bonaparte but also during the reign of his namesake Napoléon III in the Second Empire.

piece of eight

a coin of colonial-period America, both North and South, whose name comes from the anglicized form of *peso*, a word derived from the Spanish *pesa* meaning weight, and the fact that there were 8 *reales* in a *peso*. In North America it was also known as a dollar.

thaler

a silver coin issued in Germany, Austria and Switzerland. The word is short for Joachimsthaler, after Joachimsthal, a town now in the Czech Republic where the coin was first minted. The word "dollar" comes from thaler.

talent and mina

the *talent* was an ancient currency unit, and is familiar from Bible stories, but it was also used in ancient Greece. There were 60 *mina* to a *talent*. Both were also units of weight.

sequin

otherwise known as *zecchino*, a name given to various gold coins of Italian city states, Malta and Turkey. their brightness led to the current use of the word sequin to describe the tiny discs attached to cloth.

Pre-decimal British coins and notes and their nicknames

farthing	¼ penny
halfpenny ("ha'penny")	½ penny
one penny	1d
threepence ("thruppence," "threpence", "threpenny bit")	3d
sixpence ("tanner")	6d
one shilling ("bob")	1s, 1/–
florin ("two bob")	2s, 2/–
half-crown ("two and six")	2s 6d, 2/6
crown	5s, 5/–
ten-shilling note ("ten bob note")	10s, 10/–
one pound note ("quid")	£1
five-pound note ("fiver")	£5
ten-pound note ("tenner")	£10

scrip
a certificate used during share dealing, and also means the shares released under a bonus issue. The term is sometimes additionally used for paper money that is not regarded as official currency— such notes might be issued during wartime, or in conditions of hyperinflation when previously existing banknotes quickly become obsolete.

gold standard
a system of linking the value of currency to an amount of gold, once common but no longer used. A dollar or pound, for instance, had a specified value in gold. The removal of the dollar from the gold standard and subsequent devaluation led to the collapse of the Bretton Woods system of **exchange rates** in 1971–72.

exchange rate
the rate at which one currency can be traded for another. In practice, buying and selling rates are slightly different, and banks and bureaux de change generally charge a commission for a transaction. The Bretton Woods agreement controlled exchange rates in the post-Second World War period until 1972, but the widely differing strengths of world economies made the system untenable. Exchange rates are now allowed to "float," finding their own level against other currencies, though national central banks sometimes step in and either buy or sell currency to influence this level.

consumer price index
a "basket" of commonly bought items, whose changing values when added together can give an idea of changes in the cost of living for the average household in an economy. Closely linked to the inflation rate, although the latter may encompass items that do not affect all households, such as mortgage interest rates.

Treasury bill (T-bill)
a short-term bill issued by the U.S. and Canadian governments in various denominations. Treasury bills carry no interest and are tradable on a discount basis with terms to maturity of 3, 6 or 12 months. The difference between purchase price and the face amount represents the return to the investor.

Exchequer bill
an interest-earning bill issued by the British government as a way of raising funds for a contingency, such as war. Exchequer or

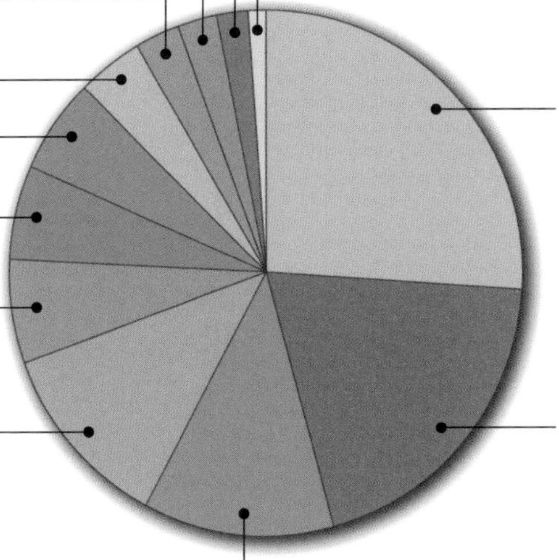

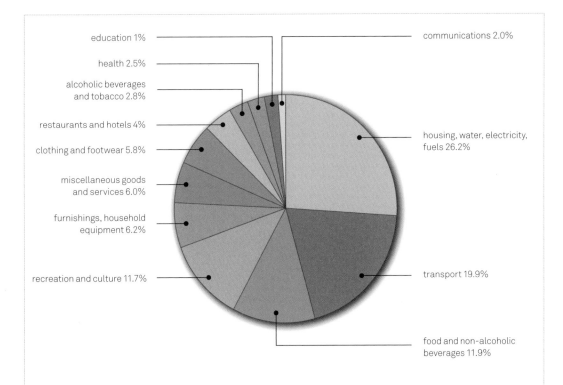

This chart shows a typical breakdown of spending for a household, changes in which indicate increases in the cost of living.

Treasury bonds for a limited period of time and at fixed interest are still often issued by governments as a way of raising funds for a specific purpose. Gilt-edged securities, or gilts, are another form of government-issued investment, considered very safe and with a fixed rate of interest.

prime rate (minimum lending rate)
the lowest rate of interest charged by major banks in the U.S., being the interest rate offered to their "prime customers." An analogous term in Britain was the minimum lending rate, though this was replaced in 1981 by the less formal "base" rate.

Gross Domestic Product (G.D.P.)
the annual total value of all goods and services created domestically within a country's economy, excluding any income from abroad.

Gross National Product (G.N.P.)
the annual total value of all goods and services created by the economy, including investment income from abroad. It is equal to Gross National Income. When calculating G.N.P., it is important to

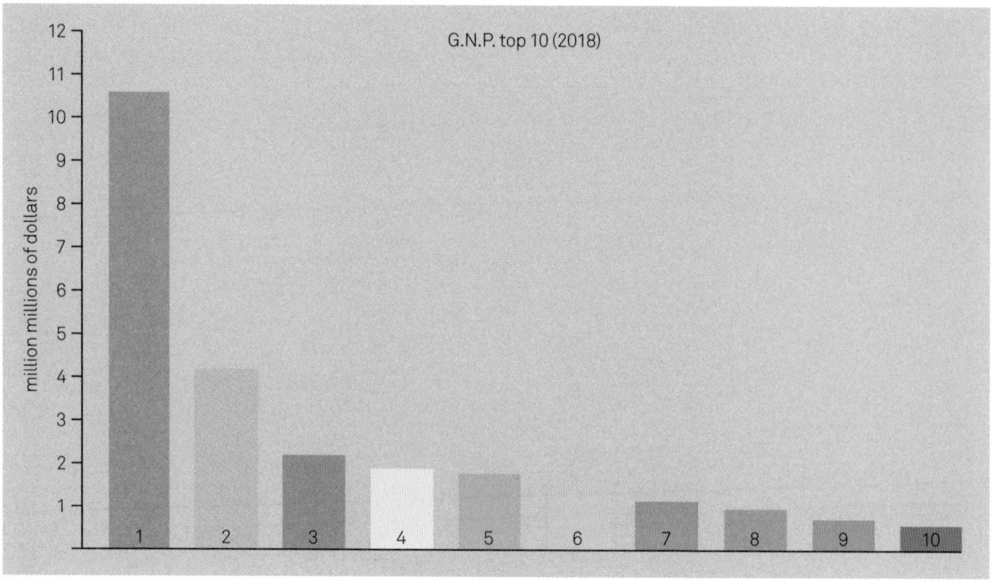

A comparison of major countries' G.N.P.s.

		Country	G.N.P. ($ mill)
●	1	United States	20,544,343
●	2	China	13,608,151
●	3	Japan	4,971,323
●	4	Germany	3,947,620
●	5	United Kingdom	2,855,296
●	6	France	2,777,535
●	7	India	2,718,732
●	8	Italy	2,083,864
●	9	Brazil	1,885,482
●	10	Canada	1,713,341

recognize the possible distorting effects of **inflation**. G.N.P. is about output not prices, and a true picture of revenue can only be arrived at by measuring costs of production as well as a rise in prices.

trade balance

the difference in value between total exports and total imports over a given period of time. Both exports and imports can be split between "visibles," i.e., tangible goods and products, and "invisibles," i.e., the output of the service sector. If total exports exceed imports, there is a trade surplus; the opposite is a deficit.

money supply

the total amount of money circulating in an economy at a particular time. Within that broad definition, money can be defined in different ways. In Britain, for instance, the two main definitions are M0—notes and coins in circulation outside the Bank of England, cash held in bank (and building society) tills, and banks' operational deposits with the Bank of England; and M4, which is M0 plus all sterling deposits held by parts of the private sector that are not banks or building societies. Control of the money supply is one tool in dealing with **inflation**.

inflation

the general rise in prices of goods and services over a fixed period of time, usually measured over a year, or the fall in the value of money over the same period. Because inflation is an average, inordinate increases in the prices of specific commodities can distort the picture, and some people's buying power can be affected much more than others. Increases in earnings also need to be taken into account to properly assess the effects of inflation. In a situation of fixed exchange rates, inflationary pressure can lead to an official devaluation of a currency.

depreciation

the fall in the value of a currency against other currencies as the result of an excess in supply of that currency. Depreciation makes imports more expensive and exports cheaper, and in theory an equilibrium will be reached with demand for the currency increasing as the country's exports become more desirable, but in reality this is often slow to take place. The term depreciation is also used in accounting when describing the fall in value of a company's equipment because of obsolescence.

duty

a **tax** on certain goods, particularly those imported into a country.

tax

a government-imposed charge, designed to raise revenue to fund spending plans. It is usually levied on income from work, sales of goods or services, etc., or based on the value of assets. Income tax is a direct tax, charged at different rates depending on the level of income; sales taxes are paid indirectly when a purchase is made.

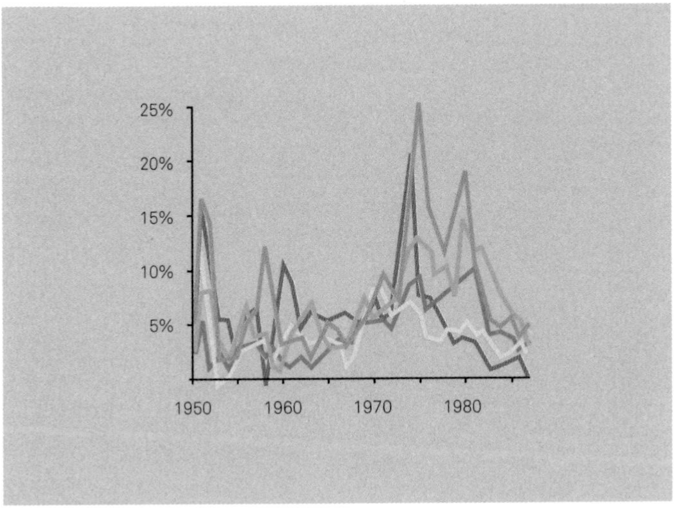

The inflation rates of these five countries varied considerably over the decades. The "oil crisis" of the 1970s affected Japan severely, but it recovered quickly.

- United States
- United Kingdom
- Japan
- Germany
- France

tithe
historically the contribution of part of the produce from farming or a portion of the income (originally a tenth part), made for the upkeep of Church and clergy in a community.

blood money (wergild)
the value placed on human life in Anglo-Saxon and early medieval Germanic communities. It amounted to compensation for the loss of life paid by the killer (whether the killing was deliberate or accidental) to the victim's family.

capital
the money that a company is able to draw on to keep it in business, or to start a business. In the latter case, it is called venture capital or risk capital, and is provided by individuals or groups in exchange for shares in the new enterprise. The term capital is also used to describe the net value of a company (after deductions have been made for taxes, overheads and wages).

interest
the charge made for borrowing money, often fixed, but sometimes varying in response to changes in the base rate, or over time. Conversely, interest is also the amount paid to investors in return for the use of their money by a government, bank or other institution at a fixed rate or again varying according to changes in the base rate.

The difference in calculation between simple and compound interest.

$$A = P \times (1 + nr)$$

simple
original amount = P
no. of years = n
amount after n years = A
annual interest rate as a proportion = r

$$A = P \times (1 + nr)^n$$

compound
A = amount of money after n years
P = principal amount
r = annual rate of interest
n = no. of years amount is deposited

Annual Percentage Rate (A.P.R.)
the annualized version of an **interest** rate expressed as a rate for a shorter period of time, e.g., per month. A monthly compound interest rate of 2 percent equals an A.P.R. of 26.24 percent. A simple A.P.R. is calculated by multiplying simple monthly interest by 12.

credit rating
the perceived ability of an individual or company to repay a loan, depending on the size of the loan. It varies according to the size of the person's income or assets, and account is also taken of the person's previous record in repaying loans.

earnings
money earned by an individual as reward for work or as interest from investments. It may be stated as the gross amount (before tax deductions) or net (after tax). Earnings also means the profits of an enterprise or different sectors of the economy.

float
a small amount of cash held at the start of trading in a shop till to provide change for initial customers, and hence to be deducted before takings are added up at the end of the day. Also used in the U.S. and Canada to mean minor cash amounts generally (often referred to as "petty cash"), or money due from uncollected checks.

share price
the price of a share in a company's equity. It is the price that a potential investor will be quoted, and as with prices of goods, is partly based on true market value, while taking into account the need to attract buyers.

yield
the return on an investment in shares or a gilt or bond. For shares, yield is measured by taking the annual **dividend** and dividing by the **share price**. A low yield may therefore indicate that the share price is high (because the market views the company's prospects favorably), or that the dividend is low because the company is performing badly.

dividend
a way of distributing part of a company's profits among shareholders, usually paid annually or half-yearly. The dividend is quoted on a per share basis.

price–earnings ratio
the ratio between the value of a share and the earnings from it, calculated by simply taking the market value of the share and dividing by the earnings per share.

turnover
the total income of an enterprise, usually measured annually, and without deducting outgoings on overheads or investment.

profit
the difference between the amount of **turnover** of an enterprise and the costs of producing its products or providing a service. Gross profit takes no account of overheads, depreciation, the amount paid in wages or **interest** payments, whereas net profit, after these deductions, is the figure most often quoted as a firm's profit (before or after **tax**). Operating profit is a firm's profit from normal trading activities, and is calculated by subtracting direct and indirect costs from the trading profit.

loss
negative profit, when the amount of **turnover** of an enterprise is outweighed by the costs of producing its products or providing its services.

margin
the **profit** margin of an enterprise is its operating profit expressed as a percentage of its turnover. For example, a company with a turnover of $4 million and an operating profit of $500,000 would have a profit margin of 12.5 percent. The operating margin is the turnover of the enterprise minus direct costs and overheads.

unit cost
the cost of producing a single item for sale, measured by taking the total cost of the production run and dividing by the number of items.

overheads
the costs of an operation that are not directly linked to the actual production of goods or services, sometimes referred to as indirect costs. These include fixed costs such as rent, maintenance contracts, and wages, as well as variable vosts such as overtime and rental of extra machinery to increase production.

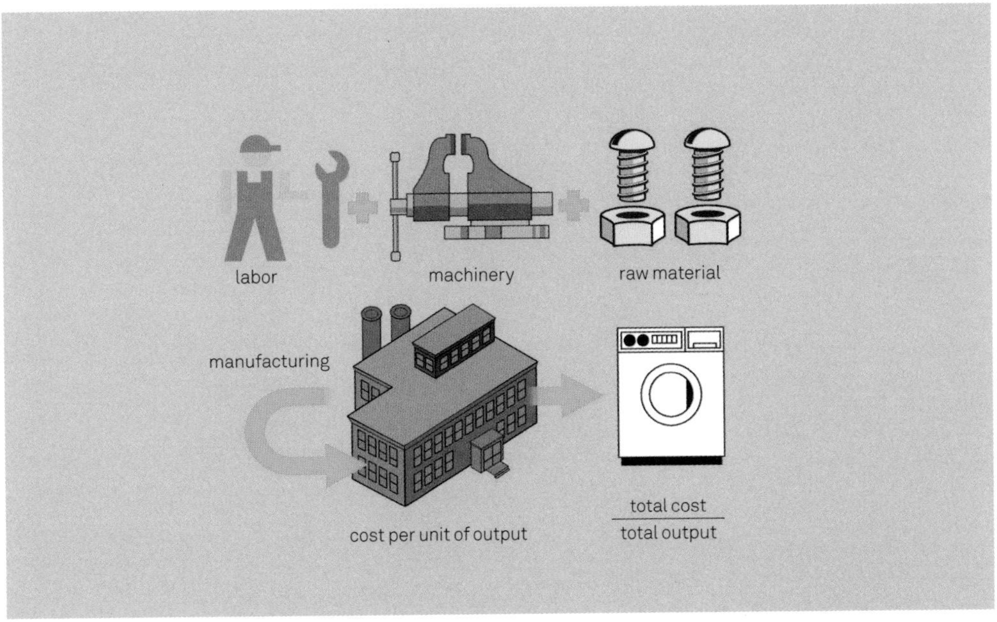

The various inputs that are brought together in the manufacturing process determine the unit cost.

working capital
the assets of a company available for use by the business in improving its turnover. Buildings that are used for production cannot be considered working capital, as they are not available as a means of raising revenue. Another definition of working capital is assets that are not offset by liabilities. In other words, assets that are a source of revenue can only be used as capital if liabilities are dealt with first.

value added
the value of an item, product, or even entire enterprise at the end of a set period minus its value at the beginning of the period. The term can also be used when describing the activity of an entire country's economy, in which case the value added amounts to the **Gross Domestic Product**. A value-added **tax** is an indirect tax that is constantly passed on through each stage of the production process. Hence, a supplier is able to deduct tax paid for materials from the tax received from a buyer.

book value
the value of a company's assets as shown in its published accounts or "books"—or the real value of a company calculated from the value of its book assets minus its book liabilities. Generally, book value is the very basic value of a company, and its "real" value as far as investors are concerned is often much higher.

Net Asset Value (N.A.V.)
the total assets of an enterprise after its liabilities and capital charges have been paid for. In this definition, it is therefore essentially the same as **book value**. In the U.S., the N.A.V. is the value of one mutual fund share, which is calculated by dividing the net assets of the fund by the number of shares outstanding.

liquidity
the amount of assets owned by an enterprise that can easily be converted into money (usually for **working capital**). The liquidity ratio is the ratio between liquid assets and the total assets of a bank or other financial institution. The related term "liquidation" refers to the termination of a business by realizing the value of assets in order to cover liabilities.

bankruptcy
the judgment by a court that a person or company is insolvent. Any property is then transferred to a trustee, who as far as possible will settle debts using the assets of the person or company. The trustee thus acts a liquidator (see **liquidity**).

Food

There are various spoon sizes available. These are the relative sizes of a tablespoon, a dessertspoon and a teaspoon.

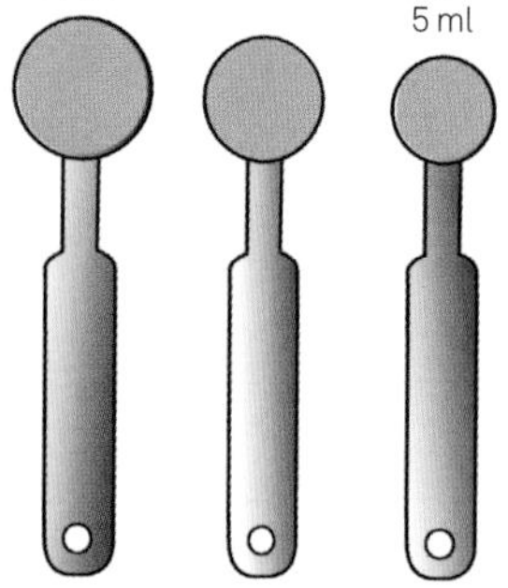

spoon sizes

measuring spoons come in various sizes, but the most common are ¼ teaspoon, ½ teaspoon, teaspoon, dessertspoon and tablespoon. Generally in liquid measures, 1 teaspoon = 5 ml (⅙ U.S. fl oz), 1 dessertspoon = 10 ml (⅓ fl oz) and 1 tablespoon = 15 ml (½ fl oz), though there is no universal agreement on these definitions. With dry measures, 1 tablespoon = ½ oz (14.235 grams).

stick

a stick of butter is equivalent to 8 tablespoons (½ cup, 4 oz, or 125 grams).

cup

for liquid measures, 1 cup = 8 U.S. fl oz (237 ml, rounded up in recipes to 250 ml) or ½ U.S. pint. In North America, a cup is the equvalent of 250 ml for dry ingredients.

dash

for liquids, a dash is defined as 6 drops, 76 drops making 1 teaspoon or ⅙ U.S. fl oz (5 ml), so it's a very small amount. But it can also refer to a small quantity of something dry that has a powerful flavor, like spices.

pinch

a pinch is a small amount of a dry ingredient. Common sense suggests that a pinch is literally just that: no more than you can get between thumb and forefinger, sometimes defined as ⅛ teaspoon or less.

milk and cream types

there are three main types of milk: whole milk; skimmed, skim or non-fat milk; and semi-skimmed or low-fat milk. Homogenized milk has been treated so that the cream is distributed throughout the milk; otherwise, it will separate and come to the top.

Types of milk and cream

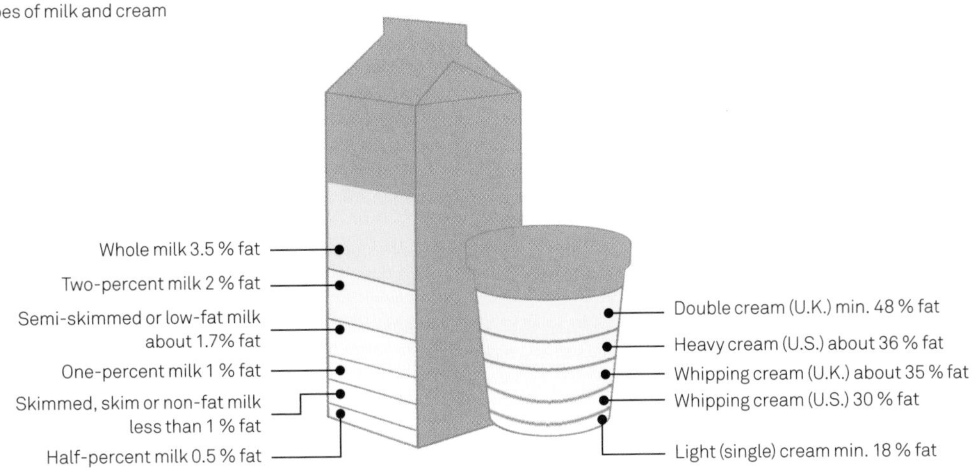

Pasteurized milk has been heat-treated to kill harmful micro-organisms. U.H.T. (ultra high temperature) pasteurization sterilizes the milk, allowing it to keep much longer, but significantly alters the taste. The main types of cream are heavy (double in the U.K.), whipping and light or single. Another variation is sour or soured cream, which has had a natural culture added to it.

egg sizes

in the U.S. size is determined by minimum weight per dozen and in Canada by weight per egg. The North American sizes are pee wee (15 oz/less than 42 grams); small (18 oz/42 grams); medium (21 oz/49 grams); large (24 oz/56 grams); extra large (27 oz/68 grams); and jumbo (30 oz/70 grams). In Britain, egg sizes are small (up to 52.9 grams); medium (53–62.9 grams); large (63–72.9 grams); and very large (73 grams and over). Eggs are also graded. In the U.S., AA are the best, followed by A, B and C. In Britain, there are two grades: A, which are sold as eggs, and B, which are broken out and pasteurized.

grades of sugar

granulated sugar is the standard form, with small white sugar crystals. Caster, fine or extra fine sugar has even smaller granules, and confectioner's or icing sugar is ground to a fine powder, used to make icing and for dusting cakes. Brown sugars come in various forms, mostly with smaller crystals than white sugar, and contain molasses. Muscovado or Barbados sugar has a strong molasses flavor and Demerara has golden, slightly sticky granules.

Sugar cooking stages. The slight differences in temperature make significant differences to the quality of syrup:

dark caramel
350–360°F/176–182°C
light caramel
320–338°F/160–170°C
hard crack
300–310°F/149–154°C
soft crack
270–290°F/132–143°C
hard ball
250–265°F/121–129°C
firm ball
242–248°F/116–120°C
soft ball
234–240°F/112–116°C
blow or souffé
230–235°F/110–112°C
thread
223–235°F/106–112°C
pearl
220–222°F/104–106°C

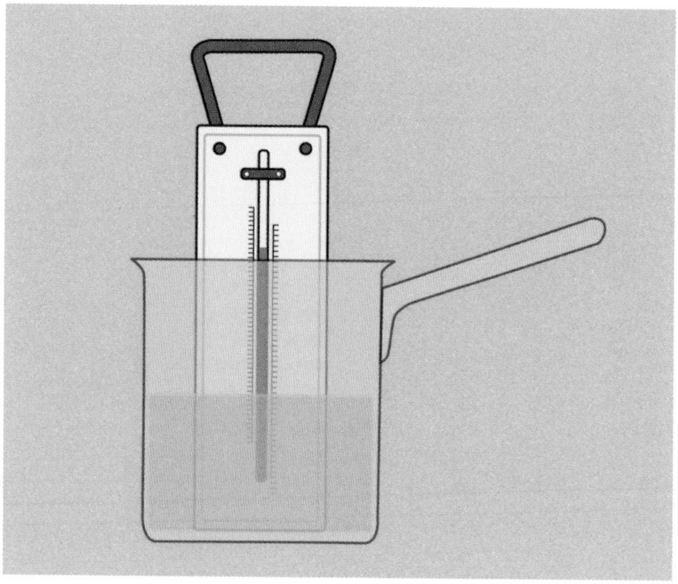

sugar cooking stages

the different stages in the cooking of sugar with water to make syrup. The higher the temperature, the more water evaporates, and the harder the syrup will set when it cools. Each stage has a name and each is separated by only a few degrees.

unit (of alcohol)

a term to describe a measure of alcohol, used particularly in health warnings. A unit is 8 grams or 10 ml of pure alcohol. The number of units in a drink is defined as its volume in milliliters multiplied by the percentage of alcohol by volume (% A.B.V.) and divided by 1,000. Thus 500 ml of beer with a 5 percent A.B.V. is 2.5 units. Safe levels of alcohol consumption are defined as a certain number of units per day, fewer for women than for men.

% A.B.V. (alcohol by volume)

the proportion of alcohol in a drink, expressed in percentage terms. One fairly reliable way of calculating it is to take a reading of the original specific gravity before fermentation using a hydrometer, measuring again once fermentation is complete. The A.B.V. is the difference between the two readings divided by 7.36. In the U.S., the A.B.W. (alcohol by weight) figure is sometimes used. To convert A.B.W. to A.B.V., multiply by 1.267.

% proof (degrees proof)

proof refers to the amount of alcohol in a drink. Despite the common use of the % sign, the percentage of alcohol is actually half the proof number. So, in the U.S., 100 proof means the drink has 50 percent alcohol. Proof was originally calculated by adding gunpowder to a drink and trying to ignite it. It was thought this would happen as soon as the percentage of alcohol reached 50 percent. The correct figure has since been found to be 57.15 percent and the British proof system is still based on this, so British 100 proof spirit is a lot stronger than its U.S. equivalent.

demijohn
a large bottle with a short narrow neck, often with handles, sometimes encased in wickerwork. These bottles were originally used to transport or store liquids but are frequently used today by home winemakers to ferment the grape juice in. A common size for winemaking is about a gallon or 4.5 liters, but demijohns can be much larger than this; some can contain as much as 20 gallons—or more.

shot
a shot is defined (vaguely) as "a small amount of drink," and usually applies to spirits or hard liquor. It is sometimes defined in the U.S. more precisely as the amount of liquor that can be held in a small glass called a jigger, i.e., 1½ fl oz (44.4 ml).

measure (of spirits)
in Britain spirits were once measured in gills (1 gill = ¼ U.K. pint). The size of a measure varied regionally, from ⅙ gill to ¼ gill, but standard measures of 25 ml and 35 ml have now been introduced (the latter becoming more common as the measure used in British pubs). Similar rules exist in other countries: in Australia, spirit measures are 15 ml, 30 ml and 60 ml, but no standard measures apply in the U.S.

split
in the U.S., a measure of carbonated water or wine, usually 6 fl oz (177 ml), sometimes a little more. A quarter-bottle of wine, 187 ml, is also often referred to as a split.

glass (of wine)
wine glasses vary greatly in size, but in U.K. pubs and restaurants a "small" glass is 125 ml, a "medium" one 175 ml and a "large" one 250 ml. Away from commercial environments, the size (and shape) of glasses will vary considerably, depending on the wine to be served in them. Red wine glasses are larger than white.

drinking stein

an earthenware beer mug, often with an elaborate design, popular in Germany. A "stein" is also used to describe a quantity of beer. The most common sizes are ½ liter and 1 liter.

Calorie

in capitalised form, the amount of a specific food able to produce 1,000 (small) calories of energy. This energy is released when the food is oxidized. The large Calorie is a unit frequently seen in diet sheets and on food nutrition labeling, but any resulting weight gain from consumption will also be inflienced by the expenditure of energy.

R.D.A.

the Recommended Daily (or sometimes Dietary) Allowance, defined as the amount of a substance, a vitamin, mineral or protein, that is suggested for daily consumption in order to maintain good health. These prescribed amounts for each substance vary depending on age, gender, and health conditions such as pregnancy.

R.D.I.

the Reference Daily Intake. A term introduced to replace **R.D.A.** on voluntary nutrition labeling in the U.S., describing the amounts of vitamins, minerals and protein contained in foods. In practice,

The Recommended Daily Allowances of vitamins and minerals. All these substances can be taken as supplements but they also naturally occur in foods.

nutrient	amount
Vitamin A	5,000 international units (IU)
Vitamin C	60 milligrams (mg)
thiamin	1.5mg
riboflavin	1.7mg
niacin	20mg
calcium	1.0 gram (g)
iron	18mg
Vitamin D	400 IU
Vitamin E	30 IU
Vitamin B6	2.0mg
folic acid	0.4mg
Vitamin B12	6 micrograms (mcg)
phosphorus	1.0g
iodine	150mcg
magnesium	400mcg
zinc	15mcg
copper	2mg
biotin	0.3mg
panothenic acid	10mg

the R.D.I. values are mostly the same as the old R.D.A. values, but the notion of "recommendation" has been removed.

Daily Reference Value

the Daily Reference Values are recommended daily maximum amounts of total fat, total carbohydrate (including fiber), protein, cholesterol, potassium and sodium, established by the U.S. Food and Drug Administration. The figures vary depending on the amount of **Calories** consumed each day.

F.C.C. unit

the Food Chemical Codex unit, used to measure the purity and effectiveness of chemicals added to foods. The F.C.C. is a set of standards prepared by the U.S. Institute of Medicine for the U.S. Food and Drug Administration. It covers such things as the amount of the enzyme lactase, well known to people who are lactose-intolerant. The F.C.C. unit is not an exact amount in milligrams, because of the different ways in which the chemicals are prepared by manufacturers. Rather, it is a measure of effectiveness, regardless of the weight of the substance.

gas mark 👁

a temperature scale commonly used in gas stoves in Britain and some Commonwealth countries, with precise Fahrenheit and Celsius equivalents.

Most British recipe books will give temperatures in gas marks as well as in degrees Fahrenheit and Celsius.

gas mark	temp (°F)	temp (°C)	description
1/4	225	110	very cool
1/2	250	130	
1	275	140	cool
2	300	150	
3	325	170	very moderate
4	350	180	moderate
5	375	190	
6	400	200	moderately hot
7	425	220	hot
8	450	230	
9	475	240	very hot

Liquids

gallon 👁
a unit of liquid measurement in the imperial system, still used in the U.K., Canada and especially the U.S., but slowly being superseded in by the liter. In British and Canadian usage, the imperial gallon represents the volume occupied by 10 lb **avoirdupois** of water, or 4.54609 liters; the U.S. gallon, however, is a smaller volume, equivalent to 3.7854 liters. Both in Britain and the U.S. the gallon can be subdivided into 8 pints, or 4 quarts, or 32 gills, but these measures necessarily have differing respective values in each system. To confuse matters further, the U.K. gallon divides into 160 U.K. fluid ounces, whereas the U.S. gallon divides into 128 U.S. fluid ounces; and though the U.K. gallon is the same in both liquid and dry measurement, the U.S. dry gallon is a separate unit, equivalent to 4.40476 liters.

Probably because they are so commonly in use for everyday purchases such as milk, beer, gasoline, etc., traditional liquid measures are only very reluctantly being replaced in the U.S. and U.K.

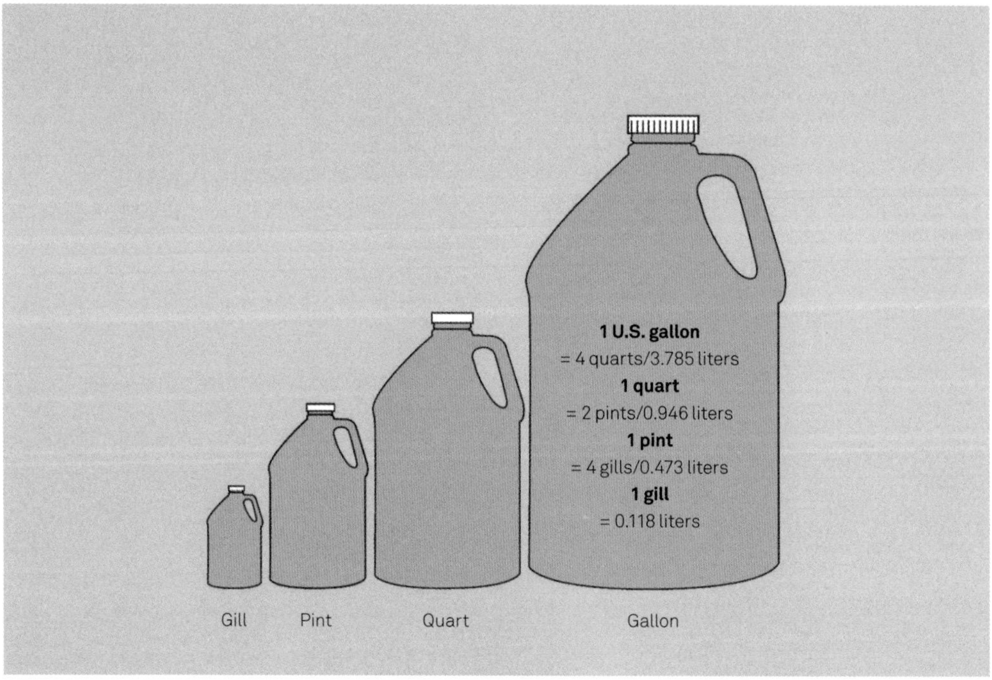

pint
a unit of liquid measurement equivalent to ⅛ gallon, ½ quart, or 4 gills. One U.K. pint is equivalent to 0.5683 liters; 1 U.S. pint equals 0.4732 liters.

fluid ounce (fl oz)
a unit of liquid measurement, a small subdivision of the pint, but representing different quantities in the U.K. and U.S. Not only is the U.K. pint a different actual volume from the U.S. pint, it is also divided into a different number of fluid ounces; whereas 1 U.K. pint is 20 U.K. fl oz (therefore 1 U.K. fl oz = 28.4131 ml), the U.S. pint is 16 U.S. fl oz (1 U.S. fl oz = 29.575 ml).

fluid dram (drachm)
a small unit of liquid measurement equivalent to ⅛ of a fluid ounce. In the U.K., this means that 1 fluid dram is 1/160 of a U.K. pint, but in the U.S. it is 1/128 of a U.S. pint—two very different quantities: in metric terms, 1 U.K. fluid dram = 3.5519 ml, 1 U.S. fluid dram = 3.6969 ml. The term fluid dram is derived from the dram or drachm of the apothecaries' weight system.

minim
a very small unit of liquid measurement equivalent to 1/60 of a fluid dram. Because of the different quantities and subdivisions of the U.K. and U.S. fluid measures, 1 U.K. minim = 0.0592 ml, and 1 U.S. minim = 0.0616 ml.

cubic meter (m^3)
a unit of volumetric measurement, normally associated with dry goods but sometimes used as a liquid measure. Unsurprisingly, it is derived from the SI meter; it is equal to the volume of a cube with sides of 1 meter, and is the equivalent of 1,000 liters or 1 kiloliter. In more common usage as a liquid measurement is the derived unit cubic centimeter (cm^3 or cc), one-millionth of a cubic meter, which is equivalent to 1 milliliter; engine capacity of motor vehicles is frequently measured in either cc or liters.

liter (litre; L; l)
a unit of volumetric measurement in the SI system, the volume of 1 kg of water at 4°C, for general purposes equivalent to 1 cubic decimeter. It is the SI unit for liquid measurement, but is also often used for dry goods. The abbreviation L was adopted officially in 1979, but l is still widely used. One liter is equal to 1.760 U.K. pints, or 2.1134 U.S. pints. Like all the SI base units,

derived units can be expressed by suffixes such as deci-, milli-, deca-, kilo-, etc.

milliliter (millilitre; mL; ml)
a unit of liquid measurement equal to one-thousandth of a liter. For general purposes, it is the equivalent of 1 cubic centimeter (cm^3 or cc), and 1 ml = 0.0352 U.K. fl oz or 0.03381 U.S. fl oz. The abbreviation mL is the officially preferred form, but ml is still widely used; in speech the term "mil" is often heard, but only in contexts where its meaning is clear, as "mil" can also be used to refer to a number of other measurements with the milli- prefix, and even as a short form of "million."

hu
a Chinese unit of liquid measurement, equal to 51.773 liters. It is subdivided into 50 *sheng*.

seah (se'a)
an ancient Hebrew unit of volumetric measurement, used for both liquid and dry goods, equivalent to approximately 13.44 liters. It was subdivided into two *hin*, and was equal to one-third of a *bath*.

kor (cor)
an ancient Hebrew unit of liquid measurement, equivalent to 10 *bath*, 30 *seah*, or about 402.3 liters. The equivalent measure for dry goods was known as a *homer* or *chomer*.

The ancient Hebrew units of liquid measure were used mainly in the oil and wine trades, but also as a measure of dry goods.

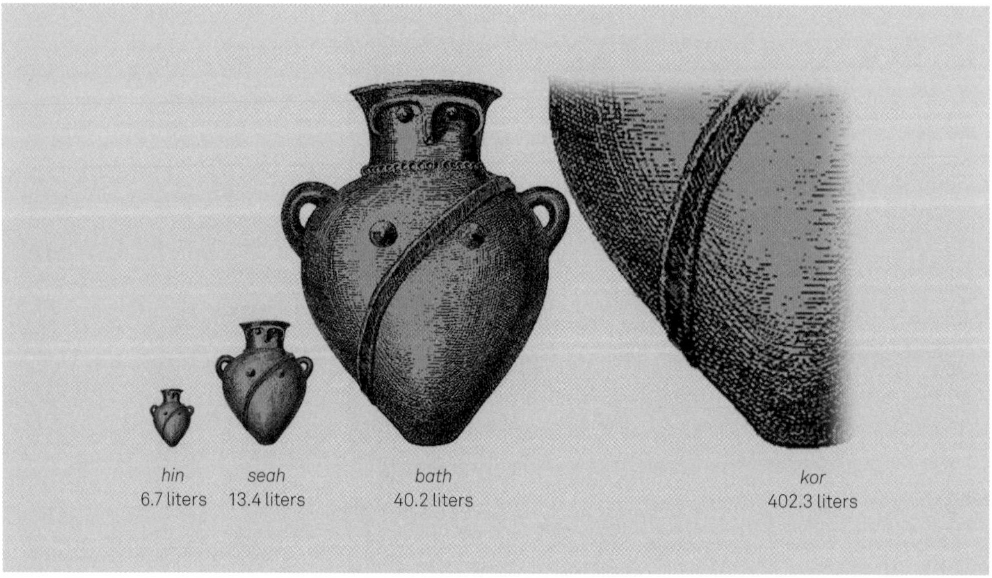

hemina
an ancient Roman unit of liquid measurement, used particularly in the wine and oil trades. It was equal to half a sextarius, and could be divided into 24 *ligulae*. It was the rough equivalent of the modern half pint, and with similar usage.

sextarius
a unit of liquid measurement common in ancient Rome, roughly equivalent to the modern pint (1 *sextarius* = 0.935 U.K. pint or 1.1227 U.S. pint), and in similar usage for wine and oil. It was equal to ⅙ of a *congius* (hence its name "*sextarius,*" Latin for "sixth"), or 2 *heminae*, approximately 0.531 liters.

congius
an ancient Roman unit of liquid measurement, equal to 6 *sextarii*, ¼ of an *urna*, or about 3.1875 liters. The term *congius* was also adopted in Britain for a time in the 19th century as an alternative name for the U.K. imperial gallon in medicine and pharmacology.

acre inch (ac in)
a volumetric unit in the measurement of bodies of water such as reservoirs. One acre inch is the volume of water covering 1 acre to a depth of 1 inch, and contains 3,630 cubic feet (approximately 102.79 cubic meters). A related, and widely used measure is the acre foot (af), equal to 12 acre inches and thus equivalent to 43,560 cubic feet (approximately 1,233.482 cubic meters).

droplet
a very small quantity of liquid, literally "a small drop." A drop was originally the unit used in dispensing and administering medicines, measured by use of a glass dropper, and roughly equivalent to a **minim**, and later taken to mean a minim. A droplet thus came to mean any quantity of liquid, smaller than a drop or minim, capable of forming a spherical shape. In practice, the term is now used for quantities smaller than about 0.05 ml, and normally measured by the diameter of the droplet rather than its volume. The size of droplets is particularly important in fields where liquids are atomized or nebulized, such as medical nasal sprays; or where the consistency of emulsions is critical, such as the food industry.

Clark scale
a scale for measuring the hardness of water, named after the 19th-century scientist Hosiah Clark. The scale is divided into units

Most hydrometers are calibrated to indicate the relative density of the liquid being tested, but some specialized instruments are graduated to show the concentration of a particular solution.

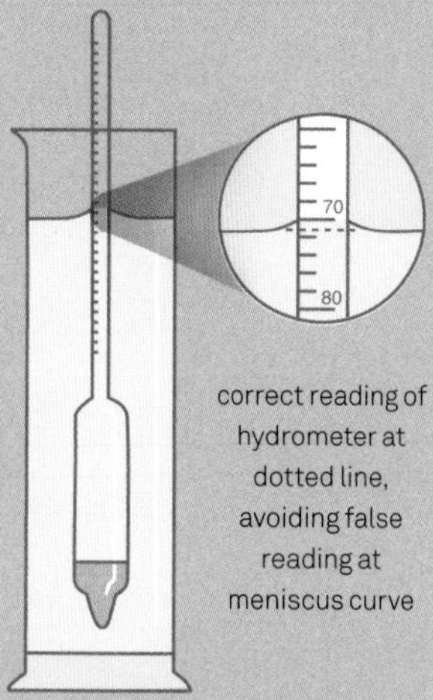

correct reading of hydrometer at dotted line, avoiding false reading at meniscus curve

known as degrees Clark: 1 degree Clark is now taken as 1 part of calcium carbonate per 70,000 of water (derived from the original unit of 1 grain per U.K. gallon), or about 14.3 parts per million.

specific gravity

the ratio of the mass of a given volume of a liquid to the mass of the same volume of water at 4°C (i.e., at maximum density). The more usual scientific term is now "relative density," but specific gravity is used (along with "original gravity," or OG, the specific gravity of the unfermented wort) in the brewing trade to measure the potential alcoholic strength of beer, although this is being superseded by reference to the percentage of alcohol in the finished product.

hydrometer

an instrument for measuring the density of liquids. It usually takes the form of a glass bulb, containing a weight, attached to a calibrated glass stem. The weight floats vertically in the liquid to be measured, and a reading can be taken at the point where the surface of the liquid reaches on the scale. A similar device, known as an oleo-meter, is used to determine the purity of oils.

hypsometer

an instrument for measuring the boiling point of liquids, usually water. As boiling point is dependent on atmospheric pressure, by comparing a hypsometer reading with boiling point at sea level, the instrument can be used to calculate altitude.

surface tension

the property of liquids that gives them the appearance of having a "skin" in a state of tension, caused by unbalanced attractive forces of molecules in the surface. Surface tension is measured in terms of the force acting on the surface, and at right angles to it, in units of force per unit length (normallynewtons per meter—Nm^{-1}). The property is responsible for such phenomena as the ability of small insects to "skate" on the surface of water, and the spherical shape of soap bubbles.

solubility

loosely, the capability of a substance to dissolve in a liquid; more specifically, the amount of a substance that can be dissolved in a liquid at a certain temperature. It can be expressed in terms of mass per unit volume, percentage, parts per million, or moles per kilogram or liter.

miscibility
the capability of two or more liquids dissolving together at a particular temperature to form a mixture.

ullage
the amount by which a container falls short of its full capacity. The term is used today mainly in shipping, and represents the spare capacity of a partially filled container (such as a tank).

sea
a large geographical mass of salt water, smaller than an ocean. The majority of the seas are in fact subdivisions of the major oceans, or immediately adjacent to them, but many, such as the Mediterranean and Baltic, are separate and virtually large salt-water lakes. Despite the old sailors' boast of having sailed "The Seven Seas," there are more than 20 bodies of water internationally recognized as seas.

ocean
a huge geographical mass of salt water. The salt water that covers more than 70 percent of the Earth's surface is normally divided into three major oceans, the Pacific, Atlantic and Indian; the Arctic and, more rarely, Antarctic Oceans are sometimes considered to be separate oceans, but geographers increasingly determine these two to be merely the northern or southern extremes of the three major oceans.

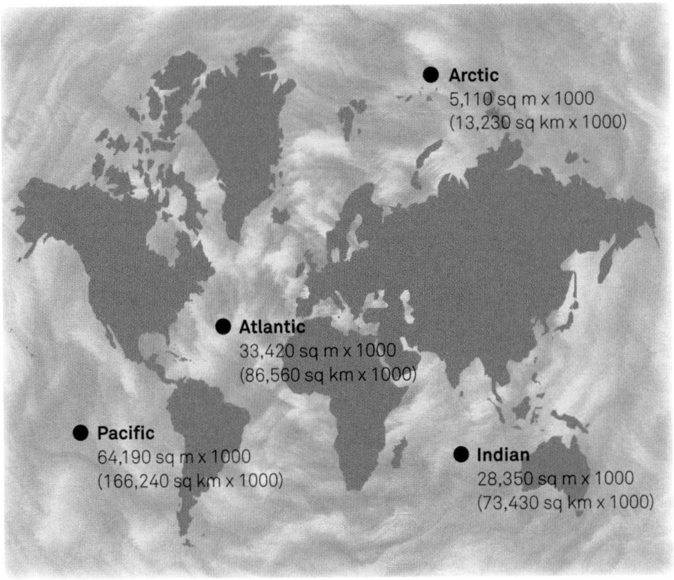

The surface area of the major oceans, excluding the adjacent seas and treating the Arctic Ocean as a separate entity, are indicated in square miles and square kilometers. The average depth of the world's oceans is around 13,100 feet (4,000 m). The greatest depth, 36,160 feet (11,022 m), is in the Pacific Ocean.

Paper and publishing

The A series of paper sizes has a constant height:width ratio, as shown above.

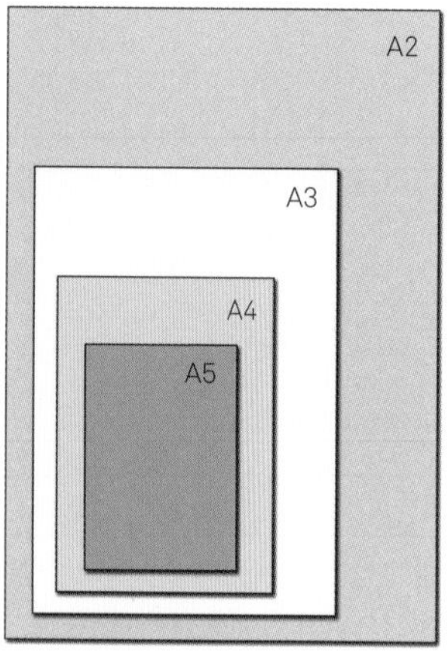

I.S.O./A.B.C. series
a system of paper sizes adopted by the I.S.O. (International Standards Organization) in Switzerland and now used by most countries outside North America. In the I.S.O./A.B.C. series, the height-to-width ratio of all pages equals the square root of two (1.4142 : 1). Two pages placed adjacently, with the longest sides touching, are equal to a larger page with the same height:width ratio, e.g., two pages of A4 equal one of A3. For some applications, the I.S.O. A series is not adequate, and the B series has been introduced to cover a wider range of paper sizes. (Japan has a slightly different set of B series sizes.) The C series defines the sizes of envelopes to fit A-series sizes of paper.

foolscap
a size of paper, now largely obsolete, used for official documents. The size was about 13 ½ × 17 inches (34.25 × 43 cm), although foolscap folio (half the size), once commonly used in offices, was often referred to simply as "foolscap." The name comes from a watermark picture of a court jester's head originally used on this type of paper.

legal cap
writing paper in tablet form, each sheet measuring 8 ½ inches × 13 to 16 inches (21.5 × 33 to 40.5 cm), with a ruled margin. Used by attorneys in the U.S. for writing legal documents.

letter
writing paper size commonly used in North and South America, with dimensions 8½ × 11 inches (21.5 × 28 cm). It is slightly wider but not as long as its metric counterpart—A4.

atlas
a large size of writing or (more usually) drawing paper generally measuring 26 × 34 inches (66 × 86.5 cm).

imperial
a size of writing or drawing paper measuring 22 × 30 inches (56 × 76.25 cm). This was the largest size of writing paper ordinarily made.

crown
a size of paper measuring 20 × 15 inches (51 × 38 cm). Before the U.K. switched to the I.S.O. paper sizes, Metric Crown (quad) paper was introduced, measuring 1,008 × 768 mm.

demy
a size of writing or drawing paper measuring 22 ½ × 17 ½ inches (57 × 44.5 cm). Double Demy measures 35 × 23 ½ inches (89 × 59.75 cm).

basic size
the usual standard sheet size of paper. The weight in grams of metric paper is based on a basic size of 1 square meter, the area of a piece of A0. U.S. paper is measured in pounds and the size used varies between different grades.

basis weight
in North America, the designated fixed weight of a ream of paper (500 sheets), measured in pounds. Also sometimes called the "ream weight" or "substance weight (sub weight)."

G.S.M.
grams per square meter, also (and more correctly) rendered as g/m^2. This measure is used in countries using the I.S.O./A.B.C. series of paper sizes to describe different thicknesses of paper.

ream
normally 500 sheets of paper, but formerly between 480 and 516 sheets. Paper is usually sold today in reams of 500 sheets, both in the metric (I.S.O./A.B.C.) system and the North American imperial system, although tissue and wrapping paper is sometimes still sold in the U.S. in reams of 480 sheets.

quire
rarely used term denoting 24 or 25 sheets of paper of the same size and quality. A quire of writing paper had 24 sheets, a quire of printing paper 25 sheets.

folio
originally a large sheet of paper folded once to make two leaves of a book, or a book made up of such sheets. The size of a folio book

is very large, something like a "coffee table" book. However, a folio may also simply be a folded sheet of any size. Several folios sewn together make a "signature"; the signatures are then bound together to make the book. Furthermore a sheet of paper in a book numbered on one side only is also referred to as a "folio," and the term is sometimes even used simply to denote a page number.

recto
the right-hand page in an open book. Thus the first page of a book is always a recto page, as are all subsequent odd-numbered pages. From the Latin word *rectus*, meaning "right."

verso
the left-hand page in an open book, the other side of a leaf from a recto page. All even-numbered pages in a book are verso pages. From the Latin word *vertere*, "to turn," because you turn the leaf to see the verso page.

-mo
a suffix forming nouns referring to the size of a book. This originally related to the amount of times a sheet of paper measuring 19 × 25 inches had been folded, although in practice several different sizes of broadsheet were used. The first three sizes were folio (folded once), quarto or 4to (twice) and octavo or 8vo (three times). A folio book had two leaves, or four pages, per sheet whereas a 64mo book had 64 leaves, or 128 pages, per sheet.

leading
the amount of space between lines of type. In the era of hand typesetting, strips of lead wee placed between the lines of type to create vertical space between the lines. "Kerning" is the horizontal equivalent in a line of type. The unit of measurement of leading is the **point**. One point is equal to 0.351 mm in the U.S. and Britain, and 0.376 mm in Europe.

point
a unit of measurement used in typography for type size, **leading** and other space specifications in a page layout. In the system used in North America and Britain there are approximately 72 points to an inch, 28 to a centimeter, and 12 points make a pica. Type size is defined as being measured from the top of the highest ascender to the bottom of the lowest descender. Different font styles mean that the actual measurement may be different from the size expressed in points.

pica
a measurement used in typography for the width of columns and other space specifications in a page layout. The etymology of the word "pica" is uncertain, although the medieval Catholic Church's rule-book on daily services was called the Pica, and there is thought to be a link. Twelve points make up a pica, and there are approximately 6 picas to an inch (just under 2½ to a centimeter). The analogous term in the **Didot point** system is the "cicero," also amounting to 12 points.

Didot point
a unit of measurement used in typography throughout Europe to measure type sizes and space specifications. There are approximately 68 Didot points to an inch (each point measures 0.376 mm). The point-based system was originally created in France by Simon Fournier, who used the French type size Cicero as the basis for the system, defining one point as a twelfth of a cicero. In 1770, François Ambroise Didot adapted Fournier's system, redefining a point as 1⁄72 of a French royal inch (slightly bigger than an imperial inch). The metric system adopted after the Revolution meant the demise of the French inch, but the Didot point system remains.

x-height 👁
the distance between top and bottom of the body of letters, in a line of type—i.e., excluding the ascenders or descenders. It is possible for letters of different typefaces to have the same **point** size but different x-heights.

To achieve an adequate amount of space between lines so that the ascenders and descenders on letters do not collide with the type on the line above or below, the type requires leading of a greater point size. The type in this caption is 7 pt, set on 10 pt leading.

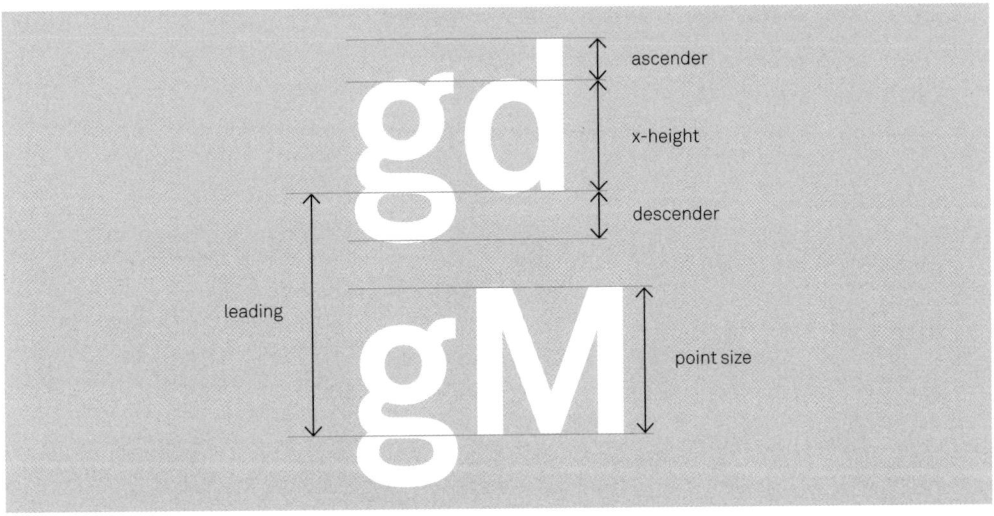

Textiles and cloth

denier 👁
a former unit for measuring the density of yarn. One denier is the density of a yarn 9,000 meters long weighing 1 gram. The denier was used especially for silk or nylon thread, and was well known as a measure of the sheerness of ladies' stockings. It has been superseded by the metric unit tex (formerly known as drex), expressed as the weight in grams of a yarn 1 kilometer long.

gram per square meter (g/m^2; gsm)
a unit of density of fabric, or more commonly, paper and board, expressed in terms of mass and surface area.

thread-count
the number of threads per unit length of a fabric. For finer fabrics this is normally expressed as threads per centimeter (or threads per inch in traditional U.K. and U.S. usage), but for coarse cloth as threads per 10 centimeters.

evenweave 👁
woven cloth in which warp and weft threads are of the same thickness and tension, and with the same number of threads per unit length and width.

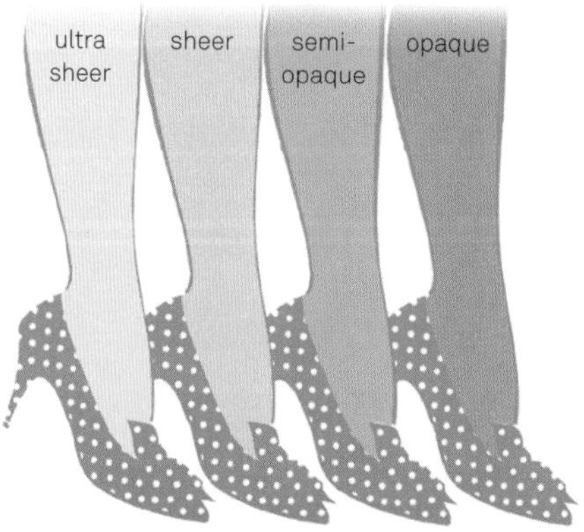

gauge
a unit of the fineness of knitted fabric. It can either refer to the number of needles per unit length of the needle bed or needle bar of the knitting machine, usually expressed as needles (or loops) per inch or per centimeter. Gauge can also refer to the size of needles used to knit a fabric of that density of knit.

ply (fold)
when applied to yarn, an indicator that the yarn is made up of more than one thread; e.g., a yarn composed of two threads twisted together would be a two-ply or two-fold yarn. The term is also used to indicate the number of folds or layers of a fabric, paper or board.

skein (lea)
a traditional unit of variable length used for measuring yarn. The value of a skein depends on the type of yarn, and can vary from manufacturer to manufacturer, but generally accepted values include:

wool and worsted	*1 skein = 80 yards (73 m)*
cotton and silk	*1 skein = 120 yards (110 m)*
linen	*1 skein = 300 yards (274 m)*

The skein is sometimes equal to $1/7$ **hank**, especially in the cotton and wool trades. In the U.S. it is more commonly called a lea.

hank
a traditional unit of variable length used for measuring yarn, its value depending on the type of yarn, and varying from one manufacturer to another. For cotton and wool, it is normally equal to seven skeins (or leas), but in the U.S. can mean 1,600 yards (1,463 m) of wool, and local usage gives several different values.

ell
a traditional unit of length used for measuring cloth. The word derives ultimately from the Latin *ulna* (elbow), which is now used as the name for a bone of the forearm, and ells were originally measured using the forearm in one way or another. As such, it was a very variable measurement, and different values evolved in different countries in Europe. In England, the ell came to be equal to 45 inches or 1.25 yards (1.143 m), whereas the Scottish ell was considerably shorter at 37.2 inches (0.945 m). The French

Different weaves used in cloth, to give different decorative or textural effects clockwise from top left: plain weave, basket weave, twill 2/2 z-wale and twill 2/2 s-wale.

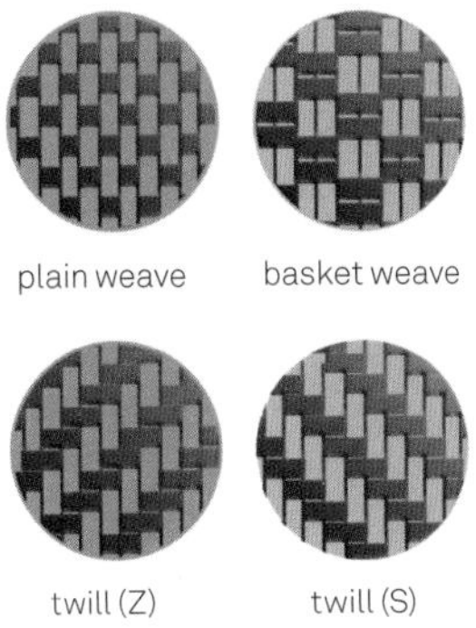

The spiral composition of the threads making up a multi-ply yarn is known as the twist, and is measured in turns per unit length (usually per inch or centimeter).

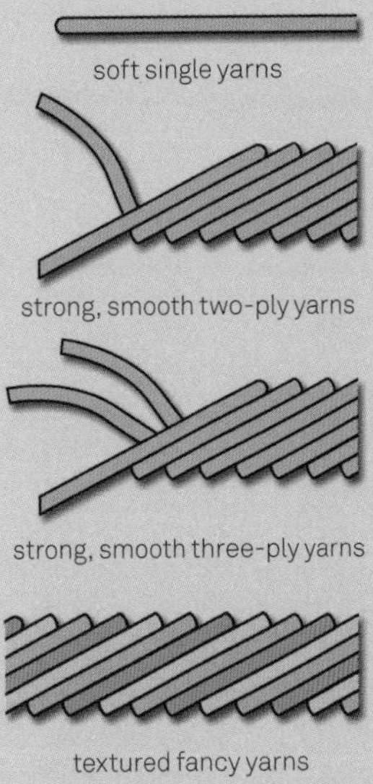

equivalent, the *aune*, was equal to 46.77 inches (1.188 m). In Germanic Europe, the related unit the *elle* was generally shorter than even the Scottish ell, being reckoned as two *fuß* (German feet), and had values ranging from about 23 to 30 inches.

bolt

a unit of length used in buying and selling cloth. Because the width of cloth varies widely depending on the type of fabric and the size of loom used in its manufacture, the actual area of a bolt varies accordingly, and even the length of a bolt is open to interpretation. Today the length of a bolt is either 30 yards (27.43 m), 40 yards (36.58 m), or 100 yards (91.44 m). It is therefore important for buyer and seller to specify which length is meant, and unsurprising that the bolt is being replaced by less ambiguous metric measurements.

width

the width of finished cloth is dependent on both the type of yarn and the size of the loom used, and is measured across the warp of the fabric. This varies widely between manufacturers, but certain

Shoe sizes remain one area where metrication has not gained a foothold—even the Continental sizing system is somewhat eccentric—and in any case most sizes are only approximate.

U.K.	U.S. ladies	U.S. men	Continental men	Mondopoint
	1			190
	1.5			
	2	1		200
1	2.5	1.5	33	
1.5	3	2	34	210
2	3.5	2.5	34	
2.5	4	3	35	
3	4.5	3.5	35	220
3.5	5	4	36	
4	5.5	4.5	37	230
4.5	6	5	38	
5	6.5	5.5	38	240
5.5	7	6	39	
6	7.5	6.5	39	250
6.5	8	7	40	
7	8.5	7.5	41	
7.5	9	8	42	260
8	9.5	8.5	42	
8.5	10	9	43	270
9	10.5	9.5	43	
9.5	11	10	44	
10	11.5	10.5	44	280
10.5	12	11	45	
11		11.5	46	290
11.5		12	47	
12				300

Men's sizes

Suits/overcoats		Shirts (collar size)		Socks	
UK/US	Continental	UK/US	Continental	UK/US	Continental
36	48	12	30-31	9.5	38-39
38	48	12.5	32	10	39-40
40	50	13	33	10.5	40-41
42	50	13.5	34-35	11	41-42
44	54	14	36	11.5	42-43
46	56	14.5	37		
		15	38		
		15.5	39-40		
		16	41		
		16.5	42		
		17	43		
		17.5	44-45		

Buying "off the peg" clothes from abroad or when traveling can be a risky business, as conversion tables often only give approximate equivalents.

traditional measurements are considered standard: for cotton the normal bolt-width is 42 inches (1.067 m), for woolen cloth 60 inches (1.524 m).

absorbency
the capacity of a fabric to absorb liquids. This can be expressed in a number of ways as the result of an absorbency test; e.g., as the change in weight of a sample of the fabric, the capillary rise of the liquid in a sample in a given time, or the time taken for a given capillary rise.

shoe sizes 👁
a unit describing (usually) the length of shoe in one of several scales in use around the world. Shoe sizes are in any case rather approximate, and depend on the size of last on which the shoe was made. The most commonly used systems are the U.S. and U.K. systems, both calibrated in increments of ⅓ of an inch (8.47 mm), and the Continental or Paris Point system, in increments of 6.66 mm. Mondopoint, a scale in millimeters to specify both length and width of foot, was adopted by the International Standards Organization, but has failed to catch on except in South Africa and parts of eastern Europe.

men's sizes 👁
traditionally, men's clothing sizes have derived from the measurements taken for bespoke tailoring, for example waist, chest, inside leg and sleeve length, in inches in the U.K. and U.S., and in centimeters in Europe. With the advent of mass production, suits and overcoats tended to be sized simply by

Buying "off the peg" clothes from abroad or when traveling can be a risky business, as conversion tables often only give approximate equivalents.

Women's sizes

Suits/dresses			Brassieres *		Hosiery	
UK	US	Continental	UK/US	Continental	UK/US	Continental
8	6	36	32 inches	70cm	8	0
10	8	38	34 inches	75cm	8.5	1
12	10	40	36 inches	80cm	9	2
14	12	42	38 inches	85cm	9.5	3
16	14	44	40 inches	90cm	10	4
18	16	46	42 inches	95cm	10.5	5
20	18	48				
22	20	50				
24	22	52				

*Note: measured across the bust for U.K. and U.S. sizes; under the bust for continental sizes

S- and Z-twist. The spinning process affects the strength of multi-ply yarns: high-twist yarns are hard and strong, suitable for weaving; low-twist yarns are soft, suitable for knitting.

chest measurement, and shirts simply by collar size, while trousers continue to be sized by both waist and inside leg measurement. Recently the pressures of mass production have led to even more standardized sizing in terms of small (S), medium (M), large (L), and extra large (XL).

dress sizes

as with **men's sizes** for clothing, the bust, waist and hip measurements used by traditional dressmakers have given way to a standardized system of sizing women's clothes.

glove sizes

gloves are traditionally sized according to the width measured around the knuckles in inches. There is a slight variation between English and continental sizing, but there is no metric equivalent to this system and it is still widely used, though it is inevitably being superseded by the ubiquitous S, M and L of mass production.

staple length

the average length of the fibers used to spin a yarn. The fibers, known as the staple, can be as short as 3 mm (⅛ inch) in cotton, and as long as 1 meter (about 39 inches) in flax. Man-made fibers are often spun into continuous filament yarn, or used as unspun monofilament, but some are cut or broken into shorter staple lengths of between 5 and 46 cm (2 and 18 inches).

twist

the number of turns per unit length of a spun yarn. To hold relatively short staple lengths of fiber in a continuous yarn, they are twisted together in the spinning process. The twist is measured in turns per centimeter (or per inch). The direction of the twist can be seen when the yarn is held vertically; if the turns are diagonal from left to right it is said to have S-twist, if from right to left, Z-twist.

hat sizes

the sizing of hats in North America is based on the circumference of the head in inches divided by pi (equivalent to the diameter if the head were a regular sphere); strangely, the U.K. sizes are equivalent to North American sizes minus ⅛. Happily, the increasingly used metric sizing system is based quite simply on the circumference of the head, although mass production is pushing the use of the even simpler S, M and L.

Hat sizes

UK	US	Metric
hat sizes		
U.K.	U.S.	metric
63/8	61/2	52
61/2	65/8	53
65/8	63/4	54
63/4	67/8	55
67/8	7	56
7	71/8	57
71/8	71/4	58
71/4	73/8	59
73/8	71/2	60
71/2	75/8	61
75/8	73/4	62
73/4	77/8	63
77/8	8	64

As with most clothing sizes, there is a bewildering difference between U.S., U.K. and metric systems. Probably the best advice is: if the cap fits ...

Music

Pitch ranges of voices

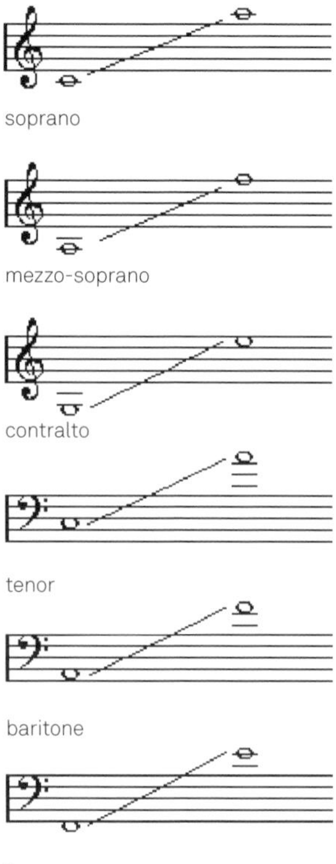

pitch range

a variety of categories of singing voice. The female soprano and unbroken male treble arre the highest, with a normal range from c′ to a″ (middle c to high a). The lower female voices are mezzo-soprano (range from a to f″, that is a below middle c to high f), and contralto (range from g to e″). Male voices range from alto (falsetto, castrato, high tenor or boy) with a normal range g to c″ through counter-tenor (the highest adult male voice, range g to e″), tenor (c to a′), and baritone (A to f′, that is the second a below middle c to the f above it), to bass (with a normal range from F to e′). A singer with a range extending below F is known as a basso profondo (deep bass). Many of the same words are used to describe musical instruments with a similar pitch range, and clef signs used in musical notation.

interval

the distance between two notes of a diatonic **scale**, described by the number of steps of the scale they include: C up to D is a second, C up to E a third, and so on. An interval also implies a ratio between the frequencies of the two pitches (e.g., octave = 1:2, perfect fifth = 2:3, perfect fourth = 3:4, major third = 4:5, minor third = 5:6), although these "pure" pitch ratios are modified in **equal temperament** to accommodate 12 exactly equal semitone steps per octave. Non-western and avant-garde music often employ intervals outside the diatonic scales, such as quarter-tones and other microtonal intervals, and cannot be described simply in conventional musical terms.

tone (whole tone; step)

the interval of a major second, unsurprisingly the sum of two semitones. A "pure" tone has the pitch ratio 8:9, but in **equal temperament** this is modified to 1:1.1224620475 (i.e., the pitch ratio of a semitone to the power of two), so that the octave is divided into six exactly equal whole-tone steps.

semitone (half-step)
the smallest interval in western diatonic and chromatic scales, the distance from one note to its immediate neighbor on the keyboard. A "pure" semitone has the pitch ratio 15:16, but in **equal temperament** this is modified to 1:1.059463094 (i.e. the 12th root of 2), which ensures a division of the octave into 12 exactly equal semitone steps.

octave
an interval in the musical scale with the frequency ratio of 1:2, comprising eight steps of a diatonic scale. Any note and its octave share the same letter name: the octave above A, for example is a, and the octave above c′ is c″. It is the only interval in **equal temperament** which is "pure," i.e., it is not altered to fit an equal division of the chromatic scale.

pitch
the standard which determines the frequency of a given note (usually a′, the A above middle C) from which all other notes of the scale can be extrapolated. The standard pitch (often called "concert pitch") of a′ = 440 Hz was established by the International Standards Organization (I.S.O.) in 1955, after centuries of confusion when the pitch of a′ varied from 400 to 500 Hz. In "authentic" performances, historic pitches such as a′ = 430 Hz and a″ = 415 Hz are still used.

The term pitch is also used loosely to assign a letter name to a particular note, and thus refers indirectly to its fundamental wave frequency. There are, confusingly, several different systems to specify which octave a particular letter-named note refers to. In the U.S. and Canada, a common usage gives middle C as C_3,

The intervals shown here are the distance from C to various other notes of the scale; similar intervals can be formed from any note.

the octave above as C_4 and the octave below as C_2, and so on. The convention common in the U.K. and Europe represents middle C as c′, the octave above as c″, the octave above that as c‴, etc. The octave below middle C is given as c, the octave below that as C, and the octave below that as C_1. In both systems pitches between these reference notes are similarly qualified—the tone above middle c′ (C_3) being d′ (D_3), the semitone below being b (B_2).

frequency
the number of vibrations per second of a musical tone, measured in hertz (Hz) or cycles per second. When referring to a complex musical note rich in overtones, the frequency of the fundamental wave determines the pitch.

scale 👁
a sequence of notes, normally spanning an octave, in ascending or descending order of pitch. In western music the most common are the diatonic scales, especially major and minor scales, and the chromatic, pentatonic and whole-tone scales, although many others (often referred to as "modes") have been devised. The diatonic scales include the major and minor scales and the church modes, and consist of seven steps within the octave comprising each of the letter-named notes in order. A sequence of twelve semitone steps is known as a chromatic scale; a sequence of six whole-tone steps forms the whole-tone scale; and any scale of five steps within the octave is known as a pentatonic scale. In non-western music, scales are often formed using pitches outside the chromatic scale involving microtonal intervals.

All major scales follow the pattern: tone, tone, semitone, tone, tone, tone, semitone. Similarly, all other scales follow the patterns as shown below, and can start on any note.

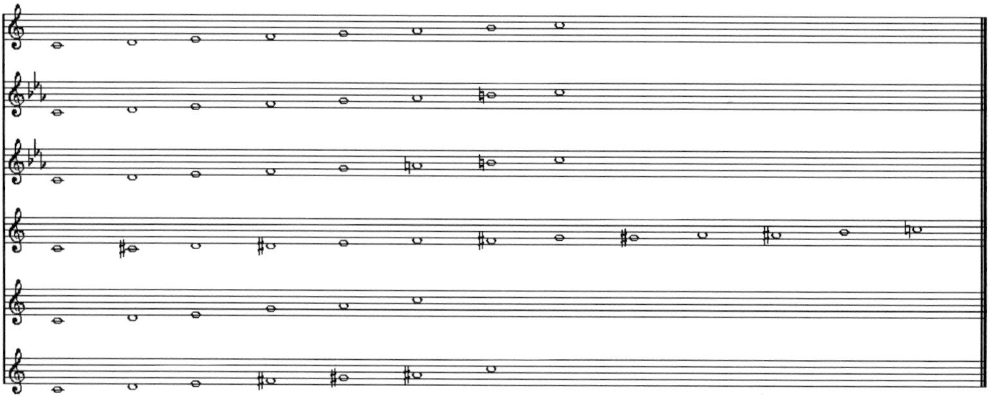

key

the diatonic scale which forms the basic tonality of a piece of music. A piece of music in a particular key uses mainly the notes of that scale and harmonies derived from those notes. The name of the key is taken from the first note, the tonic, of that scale, and though the music may move to related keys by a process of modulation, it normally starts and ends in that "home" key. Thus a piece in the key of C major would start and end with a chord of C major, and the melody and harmonies would use the notes of the scale of C major as their basis. The concept of key is not, however, universal: western music before the 16th century, non-western music, and much 20th-century avant garde music is not based on diatonic scales, so cannot be described in terms of key or tonality.

clef

a sign placed on the five lines of the musical staff to indicate the pitches of the various lines and spaces. Originally stylized versions of the letter names of the notes they denote—F, G and C—the three styles of clef in use today are, respectively:

𝄢 ***F or bass clef***

The F clef, now more commonly called the bass clef, indicates that the fourth line up represents the pitch f, and the other pitches can be deduced from this reference point.

𝄞 ***G or treble clef***

Similarly, the G clef (now usually known as the treble clef) assigns the pitch g´ to the second line up on the staff.

𝄡 ***C or alto or tenor clef***

The C clef, by contrast, can be placed in several different positions on the staff, but however placed its center determines the line assigned to the pitch c´: when it falls on the center line of the staff the clef is known as the alto clef, when on the fourth line up it is known as the tenor clef. Very rarely it is centered on other lines to become soprano, mezzo-soprano or baritone clefs.

time signature

a sign placed on the staff indicating the meter of the music, i.e., how many beats to the bar (or measure), and what kind of beat. The time signature consists of two figures, one above the other: the lower denotes the unit of measurement relative to the **whole note** (semibreve), the upper indicates how many of these units occur in the measure. Thus a time signature of 3/4 indicates that

Many instruments are grouped into "families"; the saxophone family, for instance, comprises the soprano, alto, tenor, baritone and bass saxophones.

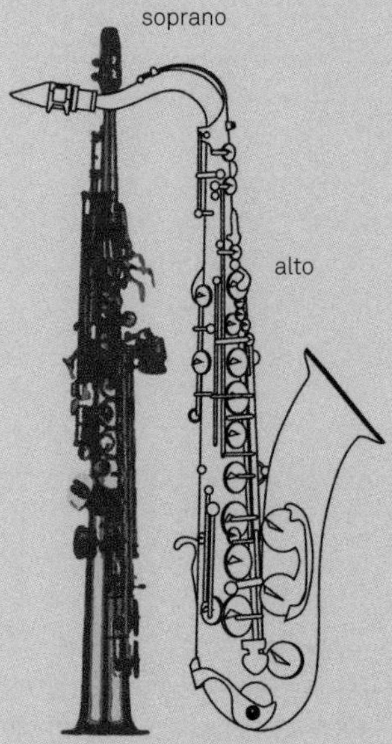

Note: not to scale

there are three **quarter-note** (crotchet) beats to the bar, 4/2 that there are four **half-note** (minim) beats to the bar.

tempo

the speed at which a piece of music is performed. Until the invention of the **metronome**, indications of tempo in musical scores were normally in the form of verbal instructions, by convention (but not exclusively) in Italian, and were subjective and relative terms giving as much an indication of the mood of a piece as its tempo. The most common of these tempo indications, in descending order of speed are:

prestissimo	extremely fast	*presto*	very fast
allegro	fast	*vivace*	lively
allegretto	quite fast	*moderato*	moderately
andante	at an easy pace	*largo*	broadly, slowish
adagio	quite slowly	*lento*	slowly
grave	seriously, slowly		

The markings may be qualified by terms such as *poco* ("a little" or "rather"), *molto* ("much" or "very"), or *ma non troppo* ("but not too much"). Changes in tempo are indicated by the markings *rit.* (ritenuto or ritardando) or *rall.* (rallentando), meaning becoming slower, and *accel.* (accelerando), meaning becoming faster.

metronome

a mechanical or electrical device for determining tempo through an audible and/or visible regular beat. A clockwork metronome was patented by J.N. Maelzel in 1815, and composers soon began to assign "M.M." numbers to their work, giving the number of beats per minute in the form M.M. = 120 (a tempo of 120 beats per minute). Later, the unit of the beat was specified in the metronome marking, for example q = 120 (a tempo of 120 crotchet beats per minute), a form which has persisted to the present.

𝅜 double note (double whole note; breve)

the longest note value in current use in western music, although its appearance is becoming rare. Its duration is equivalent to two **whole notes**, four **half-notes**, or eight **quarter-notes**, and its actual duration in time is dependent on the tempo. Paradoxically, the word breve comes from the Latin meaning "short"; originally the breve was the short note of medieval musical notation, but the longer notes of this system have fallen out of use, and shorter note values have been devised to subdivide the breve.

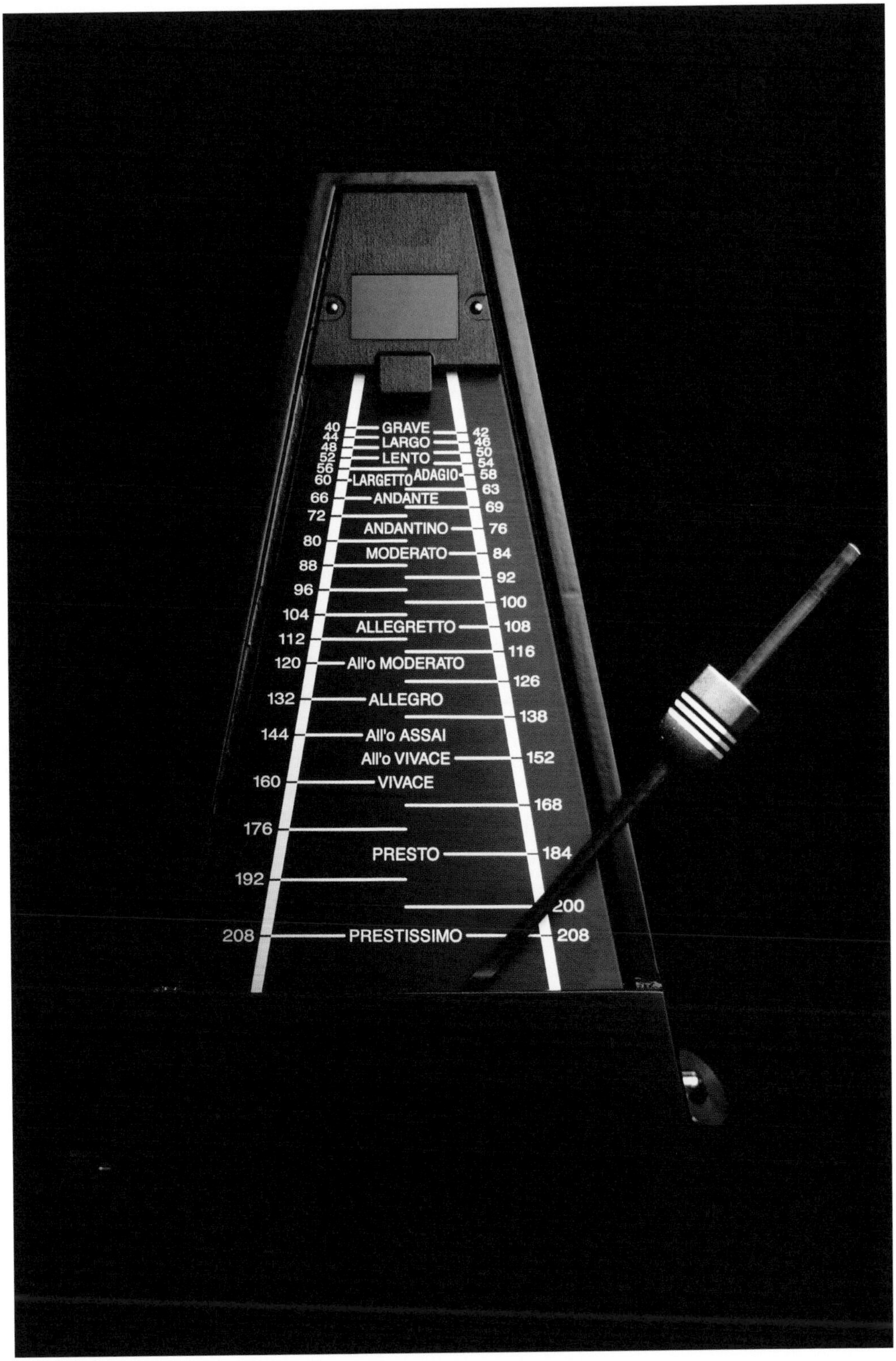
40 GRAVE 42
44 LARGO 46
48 50
52 LENTO 54
56
60 LARGETTO ADAGIO 58
66 ANDANTE 63
72 69
ANDANTINO 76
80 MODERATO 84
88 92
96 100
104 ALLEGRETTO 108
112 116
120 All'o MODERATO 126
132 ALLEGRO 138
144 All'o ASSAI
All'o VIVACE 152
160 VIVACE 168
176 PRESTO 184
192 200
208 PRESTISSIMO 208

By using dotted notes, values between those shown below can be created. A rest can also be dotted to increase its value by a half, but this is more commonly done by adding a rest of the next value down.

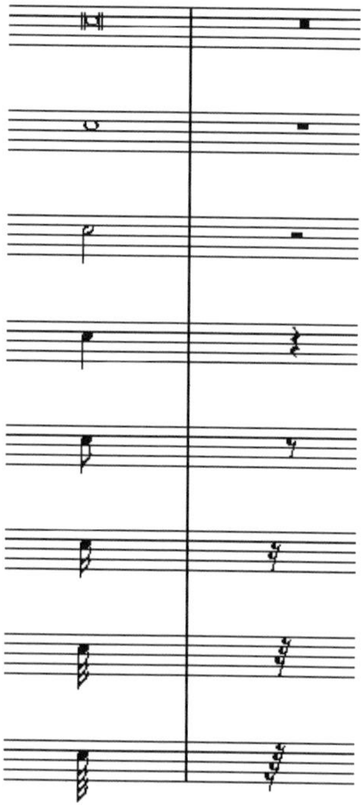

𝅝 whole note (note; semibreve)
the longest note value in common usage in contemporary music. As its name implies, its duration is equivalent to half a **double note**, two **half-notes** or four **quarter-notes**.

𝅗𝅥 half-note (minim)
a note value whose duration is, as its name implies, half that of a **whole note** (semibreve), or, following the logical sequence, the equivalent of two **quarter-notes**.

♩ quarter-note (crotchet)
a note value whose duration is, as its name implies, quarter that of a **whole note** (semibreve). In the most commonly used time signatures—for example, 4/4, 3/4 and 2/4—the measure (or bar) is divided into quarter-note beats.

♪ 1/8th note (quaver)
a note value equivalent to half a **quarter-note**, or, as its name implies, to 1/8 of a **whole note**.

𝅘𝅥𝅯 1/16th note (semiquaver)
a note value equivalent to half an **1/8th note**, or, as its name implies, to 1/16 of a **whole note**.

1/32nd note (demisemiquaver)
a note value equivalent to half a **1/16th note**, or, as its name implies, to 1/32 of a **whole note**.

1/64th note (hemidemisemiquaver)
a note value equivalent to half a **1/32nd note**, or, as its name implies, 1/64 of a **whole note**.

dotted notes
notes with a dot placed immediately after them, which increases their length by a half. A **half-note** (minim), which has a time value of two **quarter-notes** (crotchets), when dotted becomes the equivalent of three **quarter-notes**; and a dotted **quarter-note** has the time value of three **1/8th notes** (quavers). The length of a note can be further increased by the placement of an additional dot immediately after the first dot. These double-dotted notes follow a similar logic, with the second dot increasing the time value by a further quarter of the original note: a **half-note** when marked with a double dot becomes the equivalent of a **half-note** plus a **quarter-note** plus an **1/8th note**.

pianissimo
a dynamic marking (i.e., indication of loudness) in music, from the Italian meaning "very soft." It is usually abbreviated to ϖ in musical scores, and like all such dynamic markings is a relative and subjective, rather than absolute, measurement of loudness.

piano
a dynamic marking in music, from the Italian meaning "soft," usually abbreviated to ***p*** in musical scores.

mezzopiano
a dynamic marking in music, from the Italian meaning "medium soft" (somewhere between piano and mezzoforte), usually abbreviated to ***mp*** in musical scores.

mezzoforte
a dynamic marking in music, from the Italian meaning "medium loud" (not as loud as forte, but louder than mezzopiano), usually abbreviated to ***mf*** in musical scores.

forte
a dynamic marking in music, from the Italian meaning "loud," usually abbreviated to ***f*** in musical scores.

fortissimo
a dynamic marking in music, from the Italian meaning "very loud," usually abbreviated to ***f*** in musical scores.Occasionally composers will demand something even louder, indicating this with ***ff*** or ***ff***, or even further ***f***s to indicate extreme loudness.

decibel (dB)
a unit of intensity of sound, derived from the bel, but in much more common usage. Although it is a far more accurate measure of the loudness of a sound than the dynamic markings of musical notation, its use is still restricted to science rather than music. The decibel is a logarithmic unit, each 10 dB step representing a tenfold increase in sound intensity, or a doubling of the perceived loudness, on a scale starting with 0 dB as the threshold of human hearing.

bel
a unit of sound intensity, equivalent to 10 decibels. Named after Alexander Graham Bell (1847–1922).

Some familiar sounds, measured in decibels

10 dB	rustle of leaves
20 dB	library noise
30 dB	soft whisper
50 dB	light traffic noise
70 dB	vacuum cleaner/train
100 dB	thunder
130 dB	jet aircraft at take-off
180 dB	space rocket at take-off

Photography

A.S.A. /I.S.O. speed

the American Standards Association (A.S.A.) system of film speeds, now adopted by the I.S.O. (International Standards Organization). It is an arithmetic system, so 200 A.S.A. is twice as fast as 100, 400 is twice as fast as 200, etc. Despite originating in the age of photographic film, the speed measurement also proves useful for digital photography. In both case, the "speed" is a measure of how much light the film (or electronic sensor) needs to record a photographic image. For photographic films, faster speeds traditionally required coarser "grain" and produced a noisier image. In digital cameras, alteration of the speed setting modifies the gain factor used to amplify the sensor's own signal, with a similar end result; while the image is brightened, variations due to random electronic "noise" are also increased.

lag time

the time between pressing the shutter control and the actual taking of a photograph. On a conventional (film) camera, this was usually negligible. Eearly digital cameras had a significant total lag time, but this problem has been largely overcome.

For amateur photographers, a film speed of 200 I.S.O. is usually satisfactory.

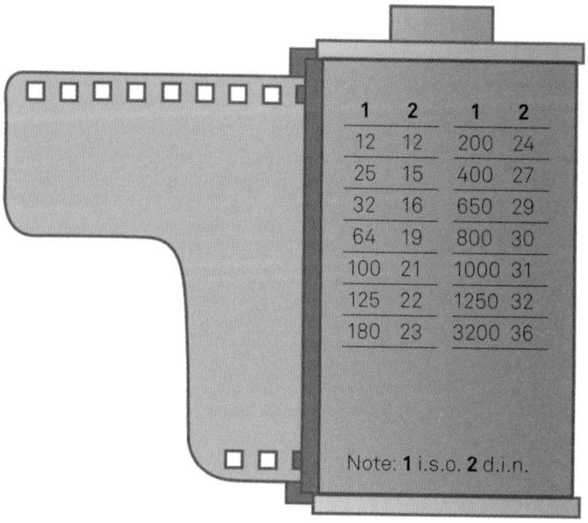
1 2 1 2
12 12 200 24
25 15 400 27
32 16 650 29
64 19 800 30
100 21 1000 31
125 22 1250 32
180 23 3200 36
Note: 1 i.s.o. 2 d.i.n.

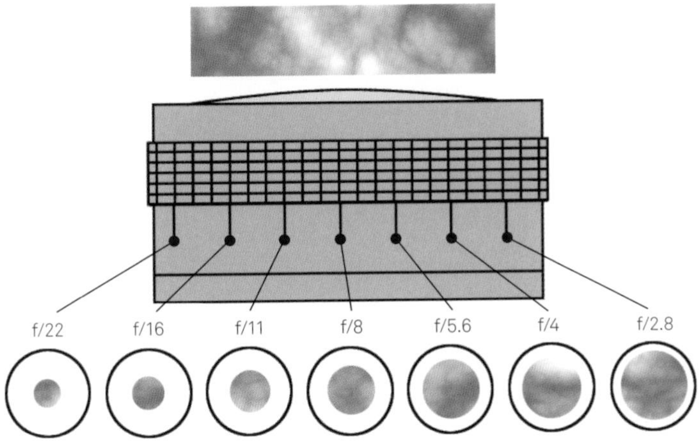

The amount of light passed through a lens is inversely proportional to the square of the f number. Modern lenses use a standard f-stop scale with the numbers in the series increasing by a factor of root 2 at each step: f/1, f/1.4, f/2, f/2.8, f/4, f/5.6, f/8, f/11, f/16, f/22, f/32, f/45 and f/64. The values of the ratios are rounded off to make them simple to write down.

shutter speed

the time the camera's shutter is held open to allow light to reach the film. Very fast shutter speeds are needed for action shots and slow speeds for night photography or special effects.

f-number

the ratio of the equivalent **focal length** to the diameter of the camera's aperture. An f number of f/1.7 means that the focal length of the lens is 1.7 times as great as the diameter. A camera with an 80-mm lens set at f/16 would mean an aperture opening of 5 mm. The lower the f-number the greater the amount of light that reaches the film.

stop

on the barrel of a camera lens, a fixed position for an **f-number** setting. Rotating the control takes it through several steps, each of which relates to a certain f-number: f/22, f/16, f/11, f/8, f/5.6, f/4, etc., each signifying a change in the aperture setting. The term "stop" is also used to measure the **dynamic range** of photographs.

exposure value (E.V.)

a value given to a combination of **shutter speed**, film speed and **f-number**. The E.V. is 0 when there is a combination of an aperture setting of f/1 with a shutter speed of 1 second at a film speed of 100 I.S.O. Each time the amount of light is halved, the E.V. increases by one. So with the same aperture setting and film speed, halving the shutter speed will increase the E.V. to 1.

gamma
a measure of contrast in a photograph relating to midtones. It is of particular interest in adjusting the contrast of digital photographs. Adjusting the contrast using photo-editing software often produces intense brightness and darkness; gamma correction is more subtle.

dynamic range
the range of light intensity levels in a photograph, from the darkest shadows to the brightest areas. Some photographic media have better dynamic ranges than others, and the same is true for digital cameras. It is measured in stops, each stop representing a doubling of the intensity of light.

color temperature
a measure of the color of light produced by a light source. The "temperature" of the light source is the temperature at which a theoretical object called a "black body" would produce the same mixture of wavelengths of light. Normally expressed in **kelvins**, color temperature for natural light typically ranges from 3,000 K (sunrise or sunset) to about 10,000 K for a heavily overcast sky. Normal daylight is around 5,000 K.

light meter
a device to measure the intensity of light from a subject, allowing camera settings to be adjusted so that the film receives the correct exposure. Also known as an exposure meter.

focal length
the distance from the center of the camera's lens to the point of focus.

Examples of color temperature. Paradoxically, the color tones that we consider "warmer"—yellows, oranges and reds—have a lower color temperature than the "colder" colors—blue through white. Candlelight and indoor electric light have much lower color temperatures than daylight. Color film is designed for daylight, and so often does not give an accurate rendering of indoor light. Tungsten film is available specifically for indoor applications.

Colour temperature

temperature (K)	light source
9,000–12,000 K	blue sky
6,500–7,500	overcast sky
5,500–5,600 K	electronic photo flash
5,500 K	sunny daylight around noon
5,000–4,500 K	Xenon lamp/light arc
3,400 K	1 hour from dusk/dawn
3,400 K	tungsten lamp
3,200 K	sunrise/sunset
3,000 K	200W incandescent lamp
2,680 K	40W incandescent lamp
1,500 K	candlelight

macro
photography of small objects, either showing them actual size or magnifying them so that they appear a lot bigger than they really are. The normal definition is a range from 1:1 to 10:1. By this method we can "see" objects in much greater detail than with the naked eye, which cannot focus when very close to an object.

telephoto
a telephoto lens has a number of elements whose effect is to make a subject appear closer to the camera. Telephoto lenses have a focal length longer than the normal 50 mm or 55 mm.

picture angle
the angle of the coverage of a lens. It is the angle formed at the center of the lens where a line drawn backward from the top left corner of the frame meets a line drawn from the bottom right corner. The longer the focal length, the narrower the picture angle.

back focus
a problem with focusing, when the actual focus is on a point behind the subject. It can mean that the subject is blurred and is often the result of failing to lock autofocus on the subject before the shot is taken. "Back focus" can also mean the distance between the back of the lens and the focal plane.

zoom (x)
a zoom lens has a variable **focal length** and yet is able to keep the subject in focus at any level of magnification. It is therefore different from a simple **telephoto** lens. Dividing the maximum focal length by the minimum gives the optical zoom or magnification factor. Thus a zoom lens that varies between 35 mm and 280 mm has an optical zoom or magnification factor of 8x.

zoom (mm)
zoom lenses have different ranges of **focal length**. A lens described as 35–280 has a focal length ranging from 35 mm to 280 mm.

megapixel (MP)
a million pixels. The resolution of pictures taken with digital cameras is measured in megapixels. The higher the number of megapixels, the higher the definition in the photograph produced, so the greater the enlargement possible. Since digital camera

Lenses with a focal length of 50 mm or 55 mm are described as normal, because they replicate a scene as it appears with the naked eye. Wide-angle lenses have a shorter focal length, and telephoto lenses have a longer focal length.

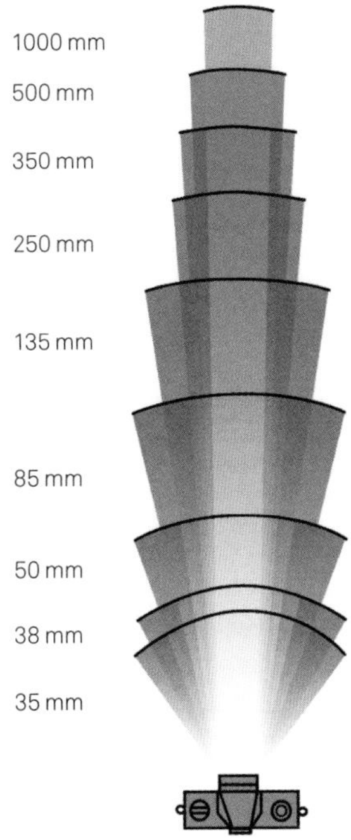

15 20 24 28 35

sensors are rectangular in shape, their resolution is equivalent to the number of pixels along the long side of the picture multiplied by the number along the short side - for example a 4,000 x 3,000 pixel sensor has a resolution of 12 MP. Note that doubling the resolution does not double the width and height: a 24 MP camera with the same sensor shape will produce images of 5657 x 4243 pixels.

effective pixels

the pixels that actually record images in a digital camera. The number of effective pixels is more important than the total of actual pixels, because the effective pixels govern the image resolution (some of the pixels are not used to record picture information). Digital cameras and phones may be advertised as offering an "interpolated" image or "digital zoom" with nearly twice as many pixels as its effective pixels. This is a software enhancement that spreads the pixels out and fills in the gaps with pixels of appropriate colors. It is less satisfactory than having a camera with that amount of effective pixels, but is a cheaper alternative.

35 mm

the most common size (width) of film, it also denotes cameras designed to take this film. The 35 mm size is felt generally adequate to provide good-quality photographs, though professional photographers may use cameras and film twice as wide. Some 35 mm cameras are of the single lens reflex variety, with a large amount of adjustment possible; others are the "point and shoot" type, which focus automatically.

Appendix 1

SI base units

Length	meter (m)
Mass	kilogram (kg)
Time	second (s)
Electric current	ampere (A)
Thermodynamic temperature	kelvin (K)
Luminous intensity	candela (cd)
Amount of substance	mole (mol)

Supplementary units

plane angle	radian (rad)
solid angle	steradian (sr)

Derived units

area	square meter
volume	cubic meter
velocity	meter per second
angular velocity	radian per second
acceleration	meter per second squared
angular acceleration	radian per second squared
frequency	hertz (Hz)
rotational frequency	reciprocal second
density and concentration	kilogram per cubic meter
momentum	kilogram meter per second
angular momentum	kilogram meter squared per second
moment of inertia	kilogram meter squared
force	newton (N)
moment of force, torque	newton meter
pressure and stress	pascal (Pa)
dynamic viscosity	pascal second
kinematic viscosity	meter squared per second
surface tension	newton per meter
energy, work, and quantity of heat	joule (J)
power and radiant flux	watt (W)
temperature	degree Celsius (°C)
thermal coefficient of linear expansion	reciprocal kelvin
heat flux density and irradiance	watt per square meter
thermal conductivity	watt per meter kelvin
coefficient of heat transfer	watt per square meter kelvin
heat capacity	joule per kelvin
specific heat capacity	joule per kilogram kelvin
entropy	joule per kelvin
specific entropy	joule per kilogram kelvin
specific energy and specific latent heat	joule per kilogram
quantity of electricity, electric charge	coulomb (C)
electric potential, potential difference, electromotive force	volt (V)
electric field strength	volt per meter
electric resistance	ohm (Ω)
electric conductance	siemens (S)
electric capacitance	farad (F)
magnetic flux	weber (Wb)
inductance	henry (H)
magnetic flux density, magnetic induction	tesla (T)
magnetic field strength	ampere per meter
luminous flux	lumen (lm)
luminance	candela per square meter
illuminance	lux (lx)
radioactivity	becquerel (Bq)
radiation absorbed dose	gray (Gy)

SI prefixes

Multiples and submultiples

yocto	y	0.000 000 000 000 000 000 000 001
zepto	z	0.000 000 000 000 000 000 001
atto	a	0.000 000 000 000 000 001
femto	f	0.000 000 000 000 001
pico	p	0.000 000 000 001
nano	n	0.000 000 001
micro	μ	0.000 001
milli	m	0.001
centi	c	0.01
deci	d	0.1
deca	da	10
hecto	h	100
kilo	k	1 000
mega	M	1 000 000
giga	G	1 000 000 000

tera	T	1 000 000 000 000
peta	P	1 000 000 000 000 000
exa	E	1 000 000 000 000 000 000
zetta	Z	1 000 000 000 000 000 000 000
yotta	Y	1 000 000 000 000 000 000 000 000

Length

picometer	pm
ångström	Å
nanometer	nm
micrometer (micron)	µm
millimeter	mm
centimeter	cm
decimeter	dc
meter	m
hectometer	hm
kilometer	km
megameter	Mm
international nautical mile	n mile (= 1 852 m)

Area

square millimeter	mm^2
square centimeter	cm^2
square decimeter	dm^2
square meter	m^2
are	a (= 100 m^2)
decare	daa
hectare	ha
square kilometer	km^2

Volume and capacity

cubic millimeter	mm^3
cubic centimeter	cm^3 (cc)
cubic decimeter	dm^2
cubic meter	m^3
cubic decameter	dam^3
cubic hectometer	hm^3
cubic kilometer	km^3
microliter (lambda)	µl
milliliter	ml
centiliter	cl
deciliter	dl
liter	l (L)
hectoliter	hl
kiloliter	kl

Mass (weight)

nanogram	ng
microgram	µg (mcg)
milligram	mg
metric carat	CM (= 200 mg)
gram	g
mounce (metric ounce)	
hectogram	hg
glug	kgm (= 0.980665 kg)
kilogram	kg
metric technical unit of mass (metric slug)	(= 9.80665 kg)
quintal	q (= 100 kg)
megagram	Mg
tonne (millier)	t

Force

micronewton	µN
dyne	dyn (= 10 µN)
millinewton	mN
pond	p (= 9.80665 mN)
centinewton	cN
crinal	(= 10 000 dyn)
newton	N
kilogram-force (kilopond)	kgf (kp) (= 9.80665 N)
kilonewton (sten, sthène)	kN
meganewton	MN

Pressure and stress

micropascal	µPa
millipascal	mPa
microbar (barye)	µbar
pascal	Pa
millibar (vac)	mbar (mb)
torr	(= ca. 133.322 Pa)
kilopascal (pièze)	kPa
technical atmosphere	at (= 9.80665 Pa)
bar	bar (b)
standard atmosphere	atm (= 101 325 Pa)
megapascal	MPa
hectobar	hbar
kilobar	kbar
gigapascal	GPa

Dynamic viscosity

centipoise	cP
poise	P (= 100 mPa s)
millipascal second	mPa s
pascal second	Pa s

Kinematic viscosity

centistokes	cSt
stokes	St (= cm^2/s)

Energy, work and quantity of heat

erg	(= 10^{-7} J)
millijoule	mJ
joule	J
kilojoule	kJ
megajoule	MJ
kilowatt hour	kWh
gigajoule	GJ
terajoule	TJ

Power

microwatt	μW
milliwatt	mW
watt	W
kilowatt	kW
megawatt	MW
gigawatt	GW
terawatt	TW
metric horsepower	ch (cv, CV, PS or pk) (= 735.498 W)

Temperature

degree Celsius	°C
kelvin	K

Electricity and magnetism

picoampere	pA
nanoampere	nA
microampere	μA
milliampere	mA
ampere	A
kiloampere	kA
picocoulomb	pC
nanocoulomb	nC
microcoulomb	μC
millicoulomb	mC
coulomb	C
kilocoulomb	kC
megacoulomb	MC
microvolt	μV
millivolt	mV
volt	V
kilovolt	kV
megavolt	MV
microhm	μΩ
milliohm	mΩ
ohm	Ω
kilohm	kΩ
megohm	MΩ
gigohm	GΩ
microsiemens	μS
millisiemens	mS
siemens	S
kilosiemens	kS
picofarad (puff)	pF
microfarad	μF
farad	F
weber	Wb
picohenry	pH
nanohenry	nH
microhenry	μH
millihenry	mH
henry	H
nanotesla	nT
microtesla	μT
millitesla	mT
tesla	T

Luminous flux

lumen	lm
lux	lx (= lm/m^2)

Radiation

becquerel	Bq
kilobecquerel	kBq
megabecquerel	MBq
gigabecquerel	GBq
gray	Gy (= J/kg)

Appendix 2

a, A — acceleration (length/time^2), atomic mass (total no. of protons and neutrons in an atom)
A — Amperes* (electric current = C/s), Angstroms* (length = 10^{-10} m), amplitude (length)
b — intercept of a linear graph, drag coefficient (mass/time)
B — magnetic field (force/current)
c — speed of light (2.998×10^8 m/s), specific heat (energy/mass × temp.), concentration (number/volume), speed of sound
cal — calories* (energy = 4.186 J)
cc — cubic centimeter
c g — group velocity
c p — phase velocity
C — Celsius* (temp.), coulombs* (electric charge), capacitance (charge/electric potential), heat capacity (energy/temp.), concentration
Cal — kilocalories* (energy)
Ci — Curie* (unit of radiation = to 3.7×10^{10} decays/s
d — distance
D — diffusion constant (area/time)
db — decibels (relative intensity)
e — electron, charge of an electron (1.602×10^{-19} C)
eV — electron Volts* (energy = 1.602×10^{-19} J)
E — energy (force × length, mass × velocity^2), electric field (force/charge)
f — frequency (1/time), focal length
f, F — force (mass × acceleration)
F — flow (volume/time), Farads* (capacitance = C/V), Fermi* (length = 10^{-15} m)
g — grams* (mass), acceleration due to gravity (9.81 m/s^2); sometimes centrifugal acceleration
G — Newton's constant (6.673×10^{-11} N m^2/kg^2), Gauss* (magnetic field = 10^{-4} T), free energy
h — height, Planck's constant (angular momentum = 6.626×10^{-34} J s), latent heat (energy/mass)
hr — hours* (time = 3600 s)
H — enthalpy (energy)
Hz — hertz (1/s)
I — moment of inertia (mass × length^2), current (charge/time), intensity (power/area), image distance
J — Joules* (energy = N m), flux (number/area time)
k — Boltzmann constant (1.381×10^{-23} J/K), spring constant (force/length), thermal conductivity (power/length temp.), wave number (1/length) (1/length)
K — kinetic energy, kelvin* (temp. = C + 273.15)
l — length, liters* (volume = 1000 cc), orbital quantum number (dimensionless, denotes angular momentum), mean free path (length)
lb — pounds* (weight; 1 kg weighs 2.2 lb)
L — angular momentum (momentum × length, moment of inertia × angular velocity)
m — mass, meters* (length), slope of a linear graph, magnetic moment (current × area), magnetic quantum number (dimensionless, denotes orientation of angular momentum)
me — mass of electron (9.109×10^{-31} kg)
mn — mass of neutron (1.675×10^{-27} kg)
mp — mass of proton (1.673×10^{-27} kg)
mi — miles* (length = 1.61 km)
min — minutes* (time = 60 s)
mmHg — millimeters mercury (pressure = 1333 dynes/cm^2)
M — molecular weight (mass/mol), magnification (dimensionless)
n — numbers of mols (dimensionless), no. of loops (dimensionless), neutron, principal quantum number (dimensionless, denotes energy level), index of refraction (dimensionless)
N — Newtons* (force = kg m/s^2), no. of particles, neutron number (no. of neutrons in an atom)
N A — Avogadro's number (dimensionless no. of objects in a mol = 6.022×10^{23})
O — object distance
p — proton
P — momentum (mass × velocity), pressure (force/area), power (energy/time)
Pa — Pascals* (pressure = N/m^2)
q, Q — charge
Q — heat (energy)
r — radius (length), distance, rate (velocity)
R — resistance (potential/current), gas constant (8.31 J/mol K)
Re — Reynolds number (dimensionless)
s — seconds*, sedimentation coefficient (time), spin quantum number (dimensionless), lens strength (1/length)
S — entropy (energy/temp.)
t — time

T	Tesla* (magnetic field = N/A m), temp.
U	potential energy (mechanical, elastic, electrical), internal energy
v	velocity (length/time), specific volume (volume/mass)
V	velocity, volume (length^3), electric potential (electric field × length), voltage, Volts* (N m/C)
W	Watts* (power = J/s), weight (force), work (energy)
x, X	horizontal position
y, Y	vertical position
z, Z	vertical position in 3D problems, atomic number (no. of protons in an atom), valence

α	alpha	angular acceleration (radians/time^2), Helium nucleus (2 p + 2 n)
β	beta	electron
Δ	delta	finite change
δ	"d"	instantaneous rate of change
ε	epsilon	electrical permittivity (e0 = 8.854 × 10^{-12} F/m), emissivity (dimensionless), efficiency (dimensionless)
φ	phi	angle
γ	gamma	electromagnetic radiation, photon
η	eta	viscosity (Poise = dyne × s/cm^2 = g/cm × s)
κ	kappa	dielectric coefficient (dimensionless)
λ	lambda	wavelength
μ	mu	magnetic permeability (m0 = 4 p × 10-7 T m/A)
ν	nu	frequency or rate of revolution (1/time)
θ	theta	angle, angular position
ρ	rho	density (mass/volume), resistivity (resistance × length)
σ	sigma	Stefan-Boltzmann constant (5.67 × 10^{-8}W/m^{2}K4)
Σ	sigma	summation
τ	tau	torque (force × length, moment of inertia × angular acceleration), radioactive half-life
ω	omega	angular velocity or angular frequency (radians/time)
Ω	omega	Ohms* (resistance = volt/ampere)

Index

Page numbers in **bold** indicate main entry.

The Quarto Group

Inspiring | Educating | Creating | Entertaining

Brimming with creative inspiration, how-to projects, and useful information to enrich your everyday life, quarto.com is a favorite destination for those pursuing their interests and passions.

First published in 2005 by Ivy Press,
An imprint of The Quarto Group
The Old Brewery, 6 Blundell Street
London N7 9BH, United Kingdom
T (0)20 7700 6700 F (0)20 7700 8066
www.quarto.com

This edition published in 2022 by Ivy Press

British Library Cataloguing-in-Publication Data
A catalogue record for this book is available from the British Library.

ISBN: 978-0-7112-6803-6
E-book ISBN: 978-0-7112-6804-3

This book was conceived, designed and produced by Ivy Press, 58 West Street, Brighton BN1 2RA, UK

Design by Glenn Howard

Printed in China
10 9 8 7 6 5 4 3 2 1

FSC www.fsc.org MIX Paper from responsible sources FSC® C016973

The publishers wish to acknowledge the following for their photographic images:

2, Shutterstock.com/Black Jack; 4, Shutterstock.com/Marina Dekhnik; 6, Shutterstock.com/Alevtina_Vyacheslav; 6, 8, 13, 24, 27, 29, 66, 117 Nova Development Corporation (Art Explosion); 7, Shutterstock.com/givaga; 25, Shutterstock.com/ Chatchawat Prasertsom; 28, Shutterstock.com/mountain beetle; 34, Shutterstock.com/Jolygon; 65, Rebecca Saraceno;84, Shutterstock.com/ Vishnevskiy Vasily; 116, Shutterstock.com/RadenVector; 133, Top left: Shutterstock.com/Anton Volynets, Top right: Shutterstock.com/ Denys Yelmanov, Bottom left: Shutterstock.com/MilonKhan, Bottom right: Shutterstock.com/S.Pytel; 172, Shutterstock.com/Gearstd; 195, Shutterstock.com/ archimede; 201, Shutterstock.com/Gts; 208, Shutterstock.com/Sergey Tinyakov; 213, David Evans.